THE BROADWAY SERIES OF ENGINEERI
VOLUME V

# STEAM TURBINES

## THEIR THEORY AND CONSTRUCTION

BY

## H. WILDA

TRANSLATED FROM THE GERMAN BY

## CHAS. SALTER

AND REVISED AND ADAPTED TO ENGLISH PRACTICE

# CONTENTS.

## CHAPTER I.

## CHAPTER II.

## THE FLOW OF STEAM THROUGH THE STEAM TURBINE.

## CHAPTER III.

### DETAILS OF STEAM TURBINES.

## CHAPTER IV.

## GENERAL ARRANGEMENT OF VARIOUS TURBINES IN PRACTICE.

## CHAPTER V.

## CONDENSERS. UTILIZATION OF EXHAUST STEAM.

## CHAPTER VI.

### PRACTICAL APPLICATION OF STEAM TURBINES.

ISBN 1-60386-034-7

# NOTATION.

The notation adopted throughout the book is in accordance with the following list. The dimensions in brackets are those adopted unless otherwise stated.

$c_p$ = specific heat of superheated steam.
$d$ = diameter of nozzle (inches).
$f$ = stress in drum due to rotation (lb. per sq. in.).
$f_1$ = stress in vanes due to rotation (lb. per sq. in.).
$g$ = acceleration due to gravity (ft. per sec.$^2$).
$n$ = number of revolutions per minute.
$p$ = pressure (lb. per sq. in.).
$r$ = relative velocity of steam to rotor vanes at outlet (ft. per sec.).
$s$ = sensible heat (B.Th.U.).
$t$ = temperature above zero (Fahrenheit).
$t_s$ = temperature of superheat (Fahrenheit).
$u$ = peripheral velocity of rotor (ft. per sec.).
$v$ = volume (cub. ft.).
$x$ = dryness fraction.

$A$ = Area (sq. in.).
$A$ = Inlet area between vanes.
$A_o$ = Outlet area between vanes.
$C_o$ = Theoretical steam consumption of turbines (lb. per hour).
$C_i$ = "Indicated" steam consumption of turbines (lb. per hour).
$C_e$ = Effective steam consumption of turbines (lb. per hour)
$D$ = Diameter of rotor (ft.).

$E$ = Energy or work (ft.-lb.).

$E_o$ = Energy or work supplied to turbine.

$E_i$ = "Indicated" or theoretical work done by turbine.

$E_L$ = Energy lost by turbine.

$H$ = Total heat of saturated steam (B.Th.U.).

$H^1$ = Total heat of superheated steam (B.Th.U.).

$I$ = Moment of inertia [in. units].

$L$ = Latent heat (B.Th.U.).

$P_e$ = Effective or Brake Horse-power.

$P^1$ = Horse-power absorbed by turbine when running "light".

$Q$ = Quantity of heat (B.Th.U.).

$R$ = Relative velocity of steam to rotor vanes at inlet (ft. per sec.).

$T$ = Absolute temperature (Fahrenheit).

$U$ = Outlet velocity of steam to rotor vanes (ft. per sec.).

$V$ = Velocity (ft. per sec.) in general and of steam at inlet to rotor vanes in particular.

$W$ = Weight (lb.).

$a$ = constant for use in formulæ for resistance to motion of rotor.

$\beta$ = constant for use in formulæ for resistance to motion of rotor.

$\gamma$ = the ratio of specific heats; the index in adiabatic expansion.

$\xi$ = coefficient used in formula for work supplied to a reaction turbine.

$\eta_i$ = "indicated" efficiency of turbine obtained from velocity diagram.

$\eta_m$ = mechanical efficiency of turbine.

$\eta_e$ = nett or overall efficiency of turbine.

$\eta_d$ = efficiency of dynamo.

$\phi$ = entropy.

$\phi_s$ = entropy of saturated steam.

$\phi^1_s$ = entropy of superheated steam

$\phi_w$ = entropy of water.

$\rho$ = density (lb. per cub. ft.).

# STEAM TURBINES, THEIR THEORY AND CONSTRUCTION.

## CHAPTER I.

### 1. *Introduction and Properties of Steam.*

(*a*) WHILST in the reciprocating engine the work done by steam in overcoming external resistance (i.e. in the performance of useful work), by the expansion of the steam at a given pressure, is due to its potential energy, in the steam turbine the kinetic energy of the steam is utilized.

In the conversion of potential into kinetic energy, the heat of steam under pressure is transformed into an equivalent amount of velocity. The pressure of the steam is totally or in part dissipated by an approximately adiabatic expansion, and is converted into velocity, in which condition the steam can do exactly the same amount of external work as corresponds to the fall of pressure. The conversion of potential into kinetic energy may be illustrated by a stretched bowstring, by means of which velocity is imparted to the arrow on being shot out.

The expenditure of power for the necessary stretching of the string, and stored up in its stretched condition, is comparable to the heat expended in giving the steam a certain pressure or power of expansion.

1

On releasing the string the stored up power is converted into kinetic energy which gives to the arrow a definite velocity, and this, when any resistance is encountered, is converted into mechanical work, by virtue of which the arrow is driven into the resisting body until the velocity imparted to it is dissipated.

(b) Adiabatic expansion results if, during the expansion of the steam, no exchange of heat and no friction takes place between the steam and the walls of the chamber in which it is expanding.

To give the steam a definite pressure, the introduction of a certain amount of heat is necessary.

This is converted into work during expansion in proportion to the external resistance, its amount corresponding with the fall of pressure, so that the amount of heat stored up in the steam after doing its work is the same as if the velocity, corresponding to the fall of pressure, had been used up in the performance of external work.

(c) For a definite fall of pressure, or the equivalent loss of velocity, there is therefore a corresponding loss of heat, so that in the reciprocating engine and in the steam turbine, the same amount of heat should theoretically be used up in the performance of equal amounts of work. The underlying principle of all types of turbines is as follows : The energy contained in steam at a given pressure is converted into velocity, and through the kinetic energy acting upon a rotor or revolving wheel by means of blades, said wheel is made to revolve. The difference in the systems of working turbines is simply whether the conversion of the fall of pressure into kinetic energy takes place, totally or partly, before the entrance of the steam into the turbine, or whether it is postponed until the turbine is reached. The conversion of the kinetic energy into mechanical work must be effected, as far as possible, to such an extent that the steam, after its passage through the turbine, only retains sufficient velocity to enable it to overcome the back pressure, either of the outer atmosphere or of the condensers.

(d) Just as work done can be expressed as the product of a force into the distance through which it acts, so can the equivalent amount of heat necessary to raise the temperature of a body, as for example water, be expressed as work, because the amount of heat necessary to raise the temperature of 1 lb. of water through 1 degree Fahrenheit or one British Thermal Unit (B.Th.U.) is equivalent to 778 foot-lbs. of work.

*Entropy.*—A change of entropy is said to occur whenever a body receives or loses heat. When such change is small the change in entropy is measured by the change of heat divided by the *absolute* temperature [461° F. below zero]. Thus if $d\phi$ is the small change in entropy, and $dQ$ is the small change in heat, and T the absolute temperature,

$$d\phi = \frac{dQ}{T} \qquad (1)$$

The entropy is usually measured from 32° F. (freezing point), and the total entropy is obtained by taking each added quantity of heat, dividing by its absolute temperature and adding together the separate results. This is expressed by

$$\phi = \Sigma \frac{dQ}{T} \qquad (1a)$$

In passing from water to steam at any temperature T, the temperature remains constant and an amount of heat equal to the latent heat L is added, so that the entropy increases by $\dfrac{L}{T}$

If therefore $\phi_s$ is the entropy of steam and $\phi_w$ is the entropy of water,

$$\phi_s = \phi_w + \frac{L}{T} \qquad (2)$$

It can be shown that the entropy of water measured from a temperature $T_0$ at absolute temperature T is given by

$$\phi_w = log_e \frac{T}{T_0} \qquad (2a)$$

For steam superheated by an amount $t_s$ we shall have entropy

$$= \phi^1{}_s = log_e \frac{T}{T_0} + \frac{L}{T} + \cdot 48\, log_e \frac{T + t_s}{T}$$

If we plot entropy as abscissæ against temperature as ordinates, then the area of a vertical strip of width $d\phi$ of the diagram will be $Td\phi = dQ$, i.e. the heat added over the portion, so that the area of a temperature-entropy diagram for a cycle of changes represents the heat added or subtracted over the cycle.

This subject of entropy frequently presents difficulty to engineers on account of the difficulty of understanding the units by which it is measured, but further treatment is, except in so far as it affects steam turbines, outside the scope of the present book.

(e) In Fig. 1 (part 1) is shown the connection

between the temperature and corresponding increase of entropy for saturated steam, the temperatures being drawn to a suitable scale as ordinates and the entropy as abscissæ. For the temperature of 356°, for example, the distance A*a* represents the entropy value consumed in heating the water from 32° to 356°, the distance *ab* giving the amount of entropy to be added for the conversion of water at 356° into dry saturated steam at 356°.

The hatched portion of the diagram thus represents a product of temperature and entropy, and therefore the amount of heat in B.Th.U. which must

## TABLE I.

### ENTROPY AND OTHER PROPERTIES OF SATURATED STEAM.

| $t°$ F. | P lb. per sq. in. (Absolute). | V. cu. ft. per lb. | Entropy | | |
|---|---|---|---|---|---|
| | | | of Water. | Increase during Evaporation. | of Saturated Steam. |
| 32 | ·089 | 3311 | ·0000 | 2·219 | 2·219 |
| 60 | ·26 | 1220 | ·055 | 2·062 | 2·117 |
| 100 | ·94 | 349 | ·130 | 1·865 | 1·995 |
| 140 | 2·88 | | ·199 | 1·694 | 1·893 |
| 180 | 7·5 | 49·7 | ·264 | 1·544 | 1·808 |
| 212 | 14·7 | 26·4 | ·313 | 1·437 | 1·750 |
| 240 | 25·0 | 16·0 | ·355 | 1·351 | 1·706 |
| 260 | 35·5 | 11·4 | ·383 | 1·294 | 1·677 |
| 280 | 49·3 | 8·4 | ·411 | 1·240 | 1·651 |
| 300 | 67·2 | 6·3 | ·438 | 1·188 | 1·626 |
| 320 | 90·0 | 4·81 | ·465 | 1·139 | 1·604 |
| 340 | 118·4 | 3·70 | ·492 | 1·092 | 1·584 |
| 360 | 153·6 | 2·90 | ·517 | 1·048 | 1·565 |
| 380 | 196·3 | 2·30 | ·542 | 1·005 | 1·547 |
| 400 | 247·8 | 1·85 | ·566 | ·965 | 1·531 |

be spent in converting saturated steam at 176° into saturated steam at 356°; or conversely, the amount of heat set free if dry saturated steam at 356° is reduced to dry saturated steam at 176°.

Should 1 lb. of steam at 356° F. be expanded adiabatically, i.e. without any addition or subtraction of heat, then the entropy remains constant; the tem-

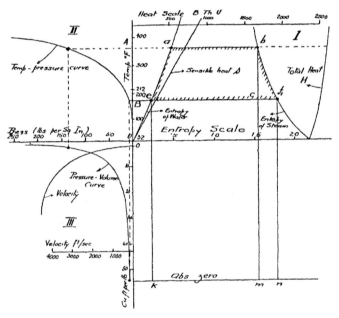

Fig. 1.—Entropy Diagram for Saturated Steam.

perature, however, falling, for example, to 176°. It is evident that steam expanded in this way cannot remain dry saturated since the entropy of such steam would be represented by the distance *ef*. Hence all the steam at the temperature 176° will be wet as there is a shortage of entropy *cf*, and of heat represented by the area *cfmn*, calculated from the absolute zero.

Consequently a partial condensation of the steam at 176° must have taken place, and the ratio of the existing condensed water to the amount of steam at 176° is expressed by the ratio, $cf : ce$.

In other words, from the original 1 lb. of steam at 356° F., there will be, as the result of the adiabatic expansion, only $\dfrac{ec}{ef}$ lb. of steam at 176° F., the remainder being water.

In the Diagram II, Fig. 1, mark off as abscissæ the values of the steam pressure in lb. per square inch corresponding to the temperatures as ordinates, and in Quadrant III mark off as ordinates the volume of steam in cubic feet per lb. corresponding to the pressure. Then the curves marked as the Temperature-Pressure and Pressure-Volume curves are obtained. From this it will be seen that for 1 lb. of saturated steam at a pressure $p = 132$ lb. per square inch, and $v = 3\cdot35$ cubic feet (Quadrant III), the corresponding temperature is $t = 348\cdot4°$ F. (T = $807\cdot8°$ F. absolute) (Quadrant II), and the entropy $1\cdot62$ (Quadrant I).

Similarly for 1 lb. of steam, $v = 54\cdot14$ cubic feet, $p = 6\cdot85$ lb. per square inch, $t = 176°$ F. The area $keabf_1n$ represents the amount of heat (in B.Th.U.) which would be necessary to transform 1 lb. of water at 32° F. absolute temperature into saturated steam at 132 lb. pressure per square inch, T = $807\cdot8°$ absolute.

If the steam is considered as brought into the condition $c$ by adiabatic expansion, and then condensed into water at absolute temperature, an amount of heat represented by the area $kef_1n$ remains, so that the amount of heat $abef_1$ has been consumed in the generation, expansion and condensation of the steam.

By saturated steam is understood that which possesses, at a given temperature, the highest possible pressure.

Should 1 lb. of saturated steam be superheated from the condition $b$, Fig. 2, through say 90° F., then

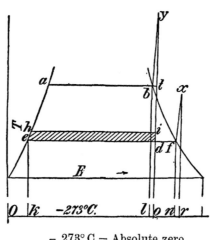

$-$ 273° C = Absolute zero.
E = Entropy.
T = Temperature.

Fig. 2.—Entropy Diagram for Superheated Steam.

heat must be introduced, and thus its temperature and entropy increased.

The increase in entropy in Fig. 2 is given by the distance $bt$, and the increase in temperature by the distance $ty = 90°$ F., so that the point $y$ represents the condition of the steam superheated by 90° F.

If the superheated steam is converted from the

condition $y$ into superheated steam in the condition $x$, its superheating being due to the friction on the sides of the vessel containing it, whilst the steam expands adiabatically, the amount of heat in 1 lb. of steam in the condition $y$ is represented by the area *keabyok*. After the expansion the amount of heat still remaining is *rxfekr*, which shows that the amount of heat corresponding to the area *eabyde* has been transformed into work, and the amount *odeko* would still remain.

If the area *abyiha* is made equal to the area *odfxro*, the area *ehide* represents the amount of heat absorbed by friction.

(*f*) If saturated steam at a pressure = $p_0$ lb. per square inch and volume = $v_0$ cubic feet per lb., is converted by adiabatic expansion into pressure = $p$ lb. per square inch and volume = $v$ cubic feet per lb., it satisfies the equation

$$p_0 v_0{}^\gamma = p v^\gamma \qquad (4)$$

In this equation, if in 1 lb. of the mixture consisting of steam and water there exists $x$ lb. in the form of steam and $(1 - x)$ lb. in liquid form, and $x > 0.7$,

$$\gamma = 1.035 + 0.1x \qquad (5)$$

For dry saturated steam, $x = 1$, therefore

$$\lambda = 1.135$$

For a mixture of 0.5 lb. water and 0.5 lb. steam = $x$ = 0.5,

$$\text{and } \gamma = 1.085$$

For superheated steam the constant $\gamma = 1.333$.
The value $x$ is known as the *dryness fraction*.

For adiabatic expansion of dry saturated steam

from $p_o$ to $p$ the dryness fraction can be calculated from the empirical formula [1]

$$x = \left(\frac{p}{p_o}\right)^{.0579} \qquad (6)$$

Further, to convert 1 lb. of water from 32° F. into steam at pressure $p$ lb. per square inch and the corresponding volume $v$ cubic feet, a total heat of H (B.Th.U.) is necessary.

In terms of the sensible heat $s$ and latent heat L

$$\text{H} = s + x\,\text{L} \qquad (7)$$

For dry saturated steam the total heat is

$$\text{H} = s + \text{L} = 1082 + 0.305t \qquad (7a)$$

when $t$ = temperature of evaporation on the Fahrenheit scale.

For superheated steam which has been obtained from saturated steam at absolute temperature $T_s$ through raising the temperature to T, we get with the specific heat at constant pressure equal to $c_p$ a total heat amounting to

$$\text{H}^1 = \text{H} + c_p\,(\text{T} - \text{T}_s) = s + \text{L} + c_p\,(\text{T} - \text{T}_s)\ \text{B.Th.U.} \qquad (8)$$

For $c_p$ an average of 0.48 may be taken.

## 2. *Flow of Steam.*

(*a*) If a quantity of steam of weight W lb. enclosed in a cylinder at a pressure = $p_o$ lb. per square inch

---

[1] Another formula which is useful and rather simpler to use is

$$\frac{x\text{L}}{\text{T}} = \frac{x_o\text{L}_o}{\text{T}_o} + \log_e \frac{\text{T}_o}{\text{T}} \qquad (6a)$$

Where T and $T_o$ are the absolute temperatures corresponding to $p$ and $p_o$ and $x$, $x_o$ are the final and initial dryness fractions. This formula is deduced from a consideration of the T-$\phi$ diagram. It is not an empirical one.—[Ed.]

and volume $= v_0$ cubic feet per lb. does mechanical work, the pressure falling adiabatically to $p$ and the volume becoming $v$, the steam then being exhausted, then the work done is represented by the hatched area in the pressure-volume diagram, Fig. 3.

Exactly the same amount of work would be given out if the steam passed from a cylinder into a space which was under the pressure $p$. If the velocity of the steam in its passage $= V$, then the steam acquires

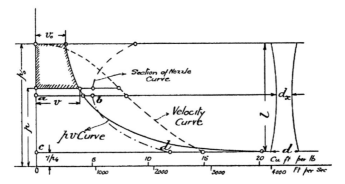

Fig. 3.—Pressure-volume Diagram, Steam Velocity and Section of Nozzle.

a kinetic energy amounting to $E =$ mass of the steam $\times \frac{1}{2}$ the square of its velocity or expressed in

$$\text{or } E = \frac{W}{g} \cdot \frac{V^2}{2} \text{ foot-lb.} \qquad (9)$$

thermal units.

$$E = \frac{1}{778} \frac{W}{g} \frac{V^2}{2} \text{ B.Th.U.} \qquad (9a)$$

As this amount of heat is necessary for the flow of the steam and for overcoming the opposing pressure $p$, the W lb. of steam at pressure $p$ and volume $v$ can only possess a total heat H which is less than the previous amount, by the amount of heat existing in the steam at pressure $= p_o$ and volume $v_o$; so that the loss of heat corresponds to the gain of kinetic energy.

Therefore for the initial condition $p_o$, $v_o$, $V_o$ we have the equation

$$E_o = \frac{1}{778} \cdot \frac{W}{g} \cdot \frac{V^2_o}{2} \text{ B.Th.U.} \qquad (10)$$

and for the final condition of the steam

$$E = \frac{1}{778} \cdot \frac{W}{g} \cdot \frac{V^2}{2} \text{ B.Th.U.} \qquad (10a)$$

Therefore the amount of diminution of heat will be

$$E_o - E = \frac{1}{778} \cdot \frac{W}{2g} (V^2_o - V^2) \text{ B.Th.U.} \qquad (11)$$

or for $W = 1$ lb. gain of K.E. $= \dfrac{1}{778} \cdot \dfrac{1}{2g} (V_o^2 - V^2)$

$$\text{B.Th.U.} \qquad (11a)$$

The greatest theoretically attainable velocity of steam in feet per second is given by the following equation :

$$V = \sqrt{2g \frac{\gamma}{\gamma - 1} p_o V_o \left[ 1 - \left(\frac{p}{p_o}\right)^{\frac{\gamma - 1}{\gamma}} \right]} \qquad (12)$$

in which $g = 32$ feet per second.[1]

Here it is assumed that the flow is through an opening whose shape permits of full development of the velocity and thus of all the kinetic energy possible in the circumstances, without loss of energy due to resistance of flow through the opening, and with no external resistance to be overcome by the expansion except the necessary amount of work for pro-

---

[1] If $H_o$ and $H_1$ (in B.Th.U.) are the total heat of the steam before and after expansion, then the velocity gain per lb. steam given by equation (11a) above $= 224 \sqrt{(H_o - H_1)}$.—[Ed.]

ducing the flow. The kinetic energy at the mouth of the opening can now be used to overcome external resistances.

The velocities of flow obtained from equation (12) in feet per second, for different values of $p_0$ and $p$, are given in Table II (p. 14), together with the difference of the corresponding total heat for dry saturated steam.

*Critical Velocity.*[1]—(b) Should the flow of steam take place through a single opening, it will be found, provided $p_0 > 2p$, that the velocity V set out in Table II will not be attained, because, on account of the resistance to the steam at the orifice, in consequence of its increasing volume, a higher pressure is set up at the orifice than exists in the vessel itself. Hence the energy of the steam is only partly converted into velocity, the amount of back pressure $p$ having very little influence.

In order to render this clear, the greatest attainable velocities V are plotted as abscissæ in Fig. 3 against the corresponding values of $p$ for a back pressure $p = 0 \cdot 1 p_0$, with $p_0 = 142 \cdot 2$ lb. per square inch, so that the curve VV represents the change of velocities.

It shows that with increasing expansion of dry saturated steam, the velocity increases also.

The assumption of adiabatic expansion appears justified by the small dimensions of the nozzle, in

[1] An examination of equation 12 for the maximum velocity per cubic feet of steam will give

$$\frac{p}{p_0} = \left(\frac{2}{\gamma + 1}\right)^{\frac{\gamma}{\gamma - 1}} \text{ which gives } V_{max.} = \sqrt{2g \frac{\gamma}{\gamma + 1} p_0 V_0}$$

These are the values referred to here as " critical ".—[Ed.]

## TABLE II.

### Velocity of Flow of Steam in Feet per Second under Varying Differences of Pressures.

| Final Pressure $p$. | Initial Pressure lb. per square inch. | | | | | | | | | | | | | | |
|---|---|---|---|---|---|---|---|---|---|---|---|---|---|---|---|
| | 14·22 | 28·44 | 42·67 | 56·89 | 71·11 | 85·34 | 99·56 | 113·78 | 128·01 | 142·23 | 156·45 | 170·67 | 184·89 | 199·12 | 213·3 |
| 1·42 | 2706 | 3083 | 3296 | 3405 | 3542 | 3609 | 3586 | 3655 | 3775. | 3831 | 3886 | 3926 | 3952 | 4018 | 4028 |
| 2·84 | 2296 | 2755 | 2984 | 3148 | 3260 | 3395 | 3460 | 3493 | 3542 | 3591 | 3673 | 3716 | 3739 | 3655 | 3805 |
| 4·26 | 2000 | 2325 | 2788 | 2952 | 3083 | 3181 | 3214 | 3312 | 3378 | 3444 | 3509 | 3542 | 3574 | 3607 | 3589 |
| 14·22 | | 1574 | 2000 | 2264 | 2427 | 2558 | 2656 | 2722 | 2821 | 2886 | 2985 | 3017 | 3050 | 3083 | 3148 |
| 28·44 | | | 1213 | 1640 | 1853 | 2033 | 2132 | 2263 | 2378 | 2460 | 2542 | 2591 | 2689 | 2702 | 2771 |
| 42·67 | | | | 1066 | 1443 | 1640 | 1804 | 1935 | 2033 | 2132 | 2230 | 2345 | 2394 | 2443 | 2510 |
| 56·88 | | | | | 885 | 1246 | 1476 | 1672 | 1771 | 1902 | 2000 | 2066 | 2148 | 2214 | 2328 |

relation to the weight of steam flowing through it.

If the value of the quotient $\dfrac{v}{V}$ be found for all values of $v$ and $V$, and these be also plotted as abscissæ to the pressure $p$, a curve is obtained in which the ordinate corresponding to the smallest abscissæ is represented by

$$Oa = 0{\cdot}5774\, p_0 = p_x = 84{\cdot}12 \text{ lb. per square inch.}$$

As the volume $v$ corresponding to 1 lb. of steam per second is always equal to the product of its velocity multiplied by the cross sectional area $A$ of the nozzle, then

$$A = \frac{v}{V} = ab \quad \text{(Fig. 3)} \qquad (13)$$

Thus it follows that there must be a pressure of $0{\cdot}5774\, p_0$ at the section $ab$, if the entire energy of the issuing steam is converted into kinetic energy.

At the end of the expansion to $p = 0{\cdot}1 p_0$, the quotient $\dfrac{v}{V}$ has attained the value $cd$, so that $cd$ represents the requisite sectional area of the mouth of the nozzle.

The shape of nozzle resulting from this sectional area, when $p = 142{\cdot}2$ lb. per square inch, and $p_0 = 14{\cdot}22$ lb. per square inch, is shown in Fig. 3, where the ratio between its smallest diameter $d_x$ and its diameter $d$ at the mouth, is $1 : 1{\cdot}56$.

The values $\dfrac{v_x}{v}$, $\dfrac{A_x}{A}$, and $\dfrac{d_x}{d}$, depending only upon the ratio $p_0 : p$, it follows that the length $l$ of the nozzle increases as $p_0 : p$ increases, and that if $p \geqq 0{\cdot}5774 p_0$, a complete conversion into kinetic energy can be obtained without conical nozzles.

Table III gives the respective values of $p_0 : p$, $d : d_x$ and $v : v_x$.

## TABLE III.

RATIO OF PRESSURE, VELOCITY, DIAMETER OF NOZZLE, AND SECTIONAL AREA AT OUTLET.

| $p_0 : p$ | 100 | 90 | 80 | 70 | 60 | 50 | 20 | 10 | 8 | 6 | 4 | 2 | 1·73 |
|---|---|---|---|---|---|---|---|---|---|---|---|---|---|
| $v : V_x$ | 2·58 | 2·56 | 2·54 | 2·51 | 2·47 | 2·43 | 2·18 | 1·92 | 1·86 | 1·74 | 1·55 | 1·02 | 1 |
| $d : d_x$ | 3·72 | 3·56 | 3·40 | 3·22 | 3·03 | 2·83 | 1·99 | 1·56 | 1·44 | 1·31 | 1·16 | 1·01 | 1 |
| $A_t : A_x$ | 13·80 | 12·69 | 11·55 | 10·39 | 9·16 | 7·98 | 3·97 | 2·44 | 2·07 | 1·72 | 1·35 | 1·02 | 1 |

If W lb. of steam flow through the nozzle per second, then equation 13 may be altered to the form

$$A = \frac{Wv}{V} \qquad (13a)$$

and the weight of steam $W_x$ flowing per second through the smallest section is determined by

$$W_x = 43 \cdot 2 \, A_x \sqrt{\frac{p_o}{v_o}} \qquad (14)$$

The velocity $V_x$ in the smallest part of the nozzle is called the critical velocity, and $p_x$ the critical pressure. These values for dry saturated steam are given by the equations

$$Vx = 70 \cdot 2 \sqrt{p_o v_o} \qquad (15)$$

and

$$p_x = 0 \cdot 5774 \, p_o \qquad (16)$$

For the inlet end of the nozzle in front of the smallest area, a careful rounding of the edges is sufficient to avoid eddy losses.

When the back pressure is low, the steam escaping adiabatically from the nozzle contains up to 20 per cent of water, which is dissipated as a cloud of vapour from the nozzle.

If the steam flowing from a nozzle is brought to rest the kinetic energy of the jet is reconverted into heat, the expanded steam becoming superheated.

If it were possible to separate the water produced during adiabatic expansion, leaving the steam alone to come to rest, the superheating would be very considerable; for example, something like 635° F. if the expanding steam had a pressure of 165 lb. per sq. inch, whereas, under the conditions given, it actually amounts to about 167° F.

The actual condition of the steam in its passage through a nozzle in which it expands adiabatically,

2

differs little from the theoretical, the chief causes of the difference being the friction that is set up and the loss of heat by conduction and radiation.

The larger the diameter of the nozzle the smaller is the ratio of circumference to area, and hence the smaller the sources of loss.

The losses in the nozzle can be taken at from 10 per cent to 15 per cent.

The volume of steam $v$ flowing through the nozzle is determined from

$$v = v_0 \left(\frac{p_0}{p}\right)^{\frac{1}{\gamma}} \qquad (17)$$

and the weight of steam W escaping per second, from

$$Wv = AV$$

where A is area in sq. inches and $v$ = volume in cubic feet per lb.

or

$$W = A_x \sqrt{2g \frac{\gamma}{\gamma + 1} \left(\frac{p_x}{p}\right)^{\frac{2}{\gamma}} \times \left(\frac{p_0}{v_0}\right)} \qquad (18)$$

If in the equations 17 and 18, for dry saturated steam, we insert $\gamma = 1{\cdot}135$, equations 14 to 16 are obtained.

From Tables II and III, as well as equations 13 to 18, the dimensions of the steam nozzle can be easily obtained.

*Example:* What must be the dimensions of a nozzle so that W = 1100 lb. of steam can pass through per hour, the steam pressure being 156 lb. per sq. inch absolute and the back pressure 2·8 lb. per sq. inch?

The critical pressure $p_x$ from equation 16 is

$$p_x = 0{\cdot}5774 \times 156 = 90 \text{ lb. per sq. inch;}$$

the greatest velocity from Table II,

V = 3673 feet per second;

the critical velocity from Table III,

$V_x = 3673 \div 2\cdot45 = 1499$ feet per second.

For $p_0 = 156$ lb. per inch we have V = 2·86 cubic feet per lb.; so from equation 14 we get

$$A_x = \frac{1100}{60 \times 60} \cdot \frac{1}{43\cdot2\sqrt{\dfrac{156}{2\cdot86}}} \times 144 = 0\cdot137 \text{ sq. inch,}$$

for which the diameter would be

$$d_x = \sqrt{\frac{4}{\pi} \times 0\cdot137} = 0\cdot414 \text{ inch.}$$

The diameter at the outflow end of the nozzle is then, from Table III,

$d = 2\cdot93 \times 0\cdot414 = 1\cdot2$ inch.

(c) A graphical determination of the nozzle section has been given by Koob, which serves its purpose in a simple and concise manner.

In Fig. 4 the three property curves are shown as in Fig. 1 in the Quadrants I, II, and III.

If in Quadrant III the curve be plotted as a velocity curve corresponding to the volume $v$, then the point $x$ is a point on the curve for the velocity V = 3340 feet per second.

If a straight line be drawn parallel to the volume axis at a distance W which corresponds to the weight of steam passing through the nozzle per second, then the line joining O$x$ cuts this parallel, and the distance of the point of intersection from the abscissæ axis represents the sectional area A, which must be provided in the nozzle in order that steam in the condition $c$, Quadrant I, when filling the section A, will flow through at the velocity V = 3840 feet.

Then from Fig. 4 we have $A : W = V : v$,

$$\text{therefore } A = \frac{Wv}{V}$$

The smallest section of the nozzle $A_x$ is given by

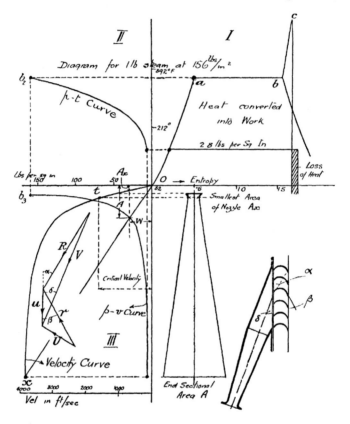

Fɪɢ. 4.—Graphical Method of Determining the Section of Nozzle.

the point of intersection of the tangent $Ot$ drawn from O to the velocity curve, and the line drawn parallel to the volume axis at the distance W.

If similar constructions be carried out for all points

of the adiabatic expansion, the diameter of the nozzle at intermediate points will be obtained.

It will be seen that in profile the nozzle is not bounded by a straight line, though the latter is constantly employed in practical work.

From this illustration it is readily seen that if the diameter of the nozzle is not altered from the point of smallest section onward, the critical velocity $V_x$ is the highest obtainable, whilst the lowest obtainable pressure is the critical one, $p_x$. By reversing the construction the changes of pressure in flowing through the nozzle can also be ascertained for any given nozzle and known initial condition of the steam.

### 3. Classification of Steam Turbines according to the Arrangement of Vanes.

The conversion of the kinetic energy of steam into useful work is effected in the steam turbine by guiding the steam on to vanes secured to a circular drum or wheel, the rotary movement produced being transmitted to the shaft of the drum or wheel [usually called the " rotor "].

(a) If the vanes are arranged radially to the axis so that the flow of steam is parallel to the axis of the turbine drum or spirally along it, then the turbine is known as an **Axial-flow Turbine.**

(b) On the other hand, when the vanes are fixed parallel to the turbine axis so that the steam flows in a radial direction, we have what are known as **Radial-flow Turbines,** wherein the steam can be directed either towards or away from the axis (**Inward and outward flow**). The manufacture of radial-flow turbines is decreasing, by far the greater number now made being axial-flow turbines.

# CHAPTER II.

## THE FLOW OF STEAM THROUGH THE STEAM TURBINE.

### 1. PERIPHERAL VELOCITY.

THE energy of steam escaping at high velocity from an orifice can be utilized economically in the steam turbine in two ways. In the one type of turbine, the **Impulse or Action Turbine,** in which the steam acts by direct impact on the vanes, the rotor, through which the energy is transmitted from the vanes to the shaft, requires to be run at a peripheral velocity, about one-half the velocity of the steam itself, and therefore about 1300 to 2000 feet per second to develop its best efficiency. In the other type—**The Reaction Turbine**—the rotors should have a peripheral velocity almost as high as the velocity of the steam ; but practical difficulties— such as stresses set up in the material by centrifugal force, in the case of large wheels even when run at relatively low speeds—prevent this ideal from being realized.

Thus, for a peripheral velocity of 1300 feet per second, if the number of revolutions should not exceed 150, the wheel diameter would be about 16 feet.

Moreover, the number of revolutions of the turbine

(22)

for direct coupled drive, must be equal to that of the driven machine; and as the highest speed used in three-phase dynamo machines hardly ever exceeds 6000 revolutions, the speed of the turbine must be reduced when using smaller turbine wheels, to correspond with the shaft speed of the direct-coupled dynamo.

**Means of reducing the number of revolutions. A. In Impulse Turbines :—**

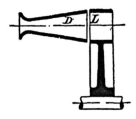
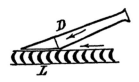

D = nozzle.                   D = nozzle.
L = rotor.                    L = rotor.

Figs. 5 and 6.—Single Stage Impulse Turbine (De Laval).

(a) The use of *speed-reducing gear*, in case the full steam velocity is to be utilized in a rotor of relatively small diameter (De Laval), Figs. 5, 6.

(b) The use of a *larger wheel diameter* without reducing gear.

(c) *Velocity Staging* (Fig. 7), in which the steam is fully expanded in the nozzles, and then enters the vanes of a rotor of such diameter that its speed corresponds to only one-half, one-third, etc., of the admis-

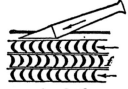

f = fixed.

Fig. 7.—Velocity Staging (Curtis Turbine).

sion velocity of the steam, so that the steam
on issuing from the rotor still retains sufficient
velocity to pass on, through fixed guide vanes,
to a second and third wheel (or more), mounted
on the same shaft, the velocity of flow being

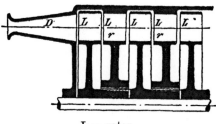

L = rotor.
L r = guide vanes.
D = nozzle.

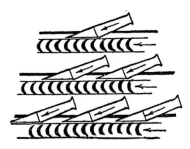

FIGS. 8 and 9.—Pressure Staging (Rateau, Zölly).

thereby gradually absorbed. In velocity staging, the pressure does not alter, but the
initial velocity is distributed, i.e. reduced in
stages.

(d) *Pressure Staging* (Figs. 8, 9) is effected
in such a manner that the total fall of
pressure takes place in several stages, the
same pressure obtaining in front and rear of
the rotor in each of the stages.

Hence, in the first rotor only part of the

pressure which existed in front of the rotor is converted into velocity through the nozzle and this velocity into work in the rotor.

The pressure still existing in. the escaping steam is again partly converted into velocity by a second nozzle apparatus, and so on until the whole fall of pressure is transformed into velocity. Hence, pressure staging corresponds to several single turbines linked up one after the other. The fall of pressure

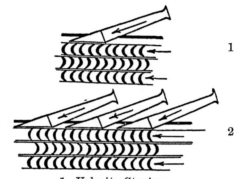

1. Velocity Staging.
2. Pressure Staging.

Fig. 10.—Pressure and Velocity Staging (Curtis).

can also be effected by modifying the sectional area of the passages in the guide vanes and rotor vanes, without the use of special nozzles.

(e) *Combination of Pressure and Velocity Staging* (Fig. 10).—Each pressure stage is divided up into velocity stages by passing the steam in one pressure stage through several rotors in succession, guide vanes being fitted between each pair of rotors, to reverse the direction of flow for the purpose of producing impact upon the next following rotor. The fall of pressure in each stage thus produces velocities which are absorbed in a number of stages.

(*f*) *The use of rotors running in opposite directions.*—The vanes of these rotors being set in opposite directions, whilst the rotor surfaces are arranged parallel to one another. The rotors are mounted close together on two shafts, so that the steam impinges upon the wheels, one after the other, each rotor thus serving as a guide wheel for one following. This system is little used in practice.

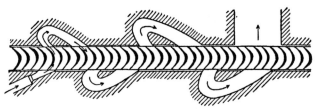

Fig. 11.—Repeated Admission (Elektra and Riedler-Stumpf Turbines).

(*g*) *Repeated admission.*—The steam (which issues from the rotor with a certain velocity) is fed through special return channels back to the rotor (Fig. 11), until the velocity of the steam is absorbed. For this purpose the return jet may be directed on to the vanes themselves or to the side faces of the rotors.

The distribution of the circulating channels over the whole circumference helps to lessen the eddy losses (see p. 38).

In a two-stage turbine of the Riedler type (Fig. 12), the steam jet is kept in a smooth curve so that the loss of energy caused by sharp bends (Fig. 11) is avoided.

**B. In Reaction Turbines** (Figs. 13, 14).—The only way in which a practicable peripheral velocity can be obtained is by multiple pressure staging, that is to say, by using numerous pressure stages, in each of which guide vanes for altering the direction of the steam are inserted between each pair of rotors. In this way, by using a sufficiently large number of pressure stages, the peripheral velocity may be reduced to any desired extent.

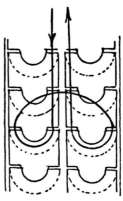

Fig. 12. — Repeated Admission (Riedler Type).

In single-stage turbines, the work must be done in about $\frac{1}{25000}$ second, and in this short time the steam has neither time nor opportunity to give up the whole of its useful energy to the rotor.

Since in multiple-pressure staging only a portion of the available heat in the steam as it issues from the nozzle is converted into kinetic energy, only a small velocity is attained, the utilization of which necessitates lower peripheral velocities than in the single-stage turbine.

By means of velocity staging the peripheral velocity can be reduced more effectually than by pressure staging.

In the latter the first rotor has to do more work than its successors, since with high steam velocities equal differences in velocity correspond to a greater capacity for doing work.

Each type of steam turbine therefore represents a compromise ; either a high velocity is used for both

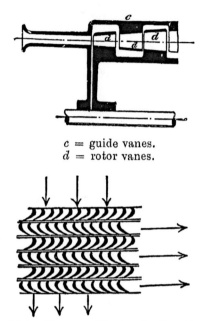

c = guide vanes.
d = rotor vanes.

FIGS. 13 and 14.—Pressure Staging for Reaction Turbines.

the steam and the rotor—in which case considerable stresses are set up in the rotor and heavy losses are incurred through friction—or a large number of stages must be employed.

### 2. CHARACTERISTIC FEATURES OF IMPULSE AND REACTION TURBINES.

#### (a) *Impulse or Action Turbines.*

In all steam turbines working by impact alone, the conversion into the kinetic energy is accomplished entirely in the nozzle, which in single-stage impulse turbines must lie in front of the rotor.

In multiple-stage turbines with velocity staging, guide channels must be fitted between the rotors, and serve simply to change the direction of flow of steam for admission to the following rotor. In those with pressure stages, nozzles—or guide vanes adapted to serve as nozzles—must be arranged between each pair of rotors, so as to provide for a further transformation into kinetic energy. The pressure of steam in the rotor is the same at the inlet and outlet. When there are so many pressure stages that the difference of steam pressure in two consecutive rotors is but small, there is no need for the guide fixed between them to be nozzle shaped, a well-rounded simple opening being sufficient.

The steam velocity for multiple-stage turbines is inversely proportional to the square root of the number of stages, provided that the same energy is to be used up in each stage. Then, if the total energy be represented by E foot-lb. per lb. of steam, the output capacity of each stage, where $m$ stages are provided, is $\dfrac{E}{m}$; and as the total loss of heat energy corresponds to the velocity,

$$V = \sqrt{2gE}$$

so we have for the velocity of a stage,

$$V_m = \sqrt{2g\,\frac{E}{m}} = \frac{V}{\sqrt{m}} \qquad (20)$$

For $V = 3280$ feet per sec. we get for 8 pressure stages the velocity at each stage $V_8 = 1160$.

Let R and $r$ represent the inlet and outlet velocity of the steam relative to the rotor, that is, the resultant of the inlet and outlet velocities of the steam and the reversed peripheral velocities of the rotor, then we have, for all pressure turbines, if $A_i$ and $A_o$ are the inlet and outlet areas between two adjacent vanes, measured perpendicular to R, $r$ and

$$A_i R \leqq A_o r \qquad (21)$$

In the Zölly steam turbine, for example, the number of stages amounts to 7 to 14 (the latter for larger powers), the steam and peripheral velocities are accordingly high. Rateau uses about 16 pressure stages.

In multiple-pressure stage turbines with multiple velocity staging in each pressure stage (Fig. 10), the relative steam velocity in the first rotor is greater than in those of only a single pressure and velocity stage, because the latter have a greater peripheral velocity.

To enable economical use to be made of the steam, the number of both pressure and velocity stages must be confined within narrow limits.

### (b) Reaction Turbines (Figs. 13, 14).

The total expansion, which is approximately adiabatic, takes place partly in the rotor and partly in the guide vanes so that the vanes of both serve as nozzles. The steam acts, as in the impulse turbine, partly by virtue of its pressure upon the vanes of the rotor and partly by the back pressure or reaction produced as it leaves the vanes; and in consequence of the expansion set up in the rotors a difference of pressure is produced at the inlet and outlet sides of the rotor, the resultant producing a thrust in the direction of the flow.

The velocity of the steam at its outlet from a reaction turbine must be less than at its outlet from an impulse turbine, energy being given up continuously. For reaction turbines the conditions always hold that

$$R > r$$
$$\text{and } RA_i > rA_o \qquad (21a)$$

Since in reaction turbines the loss of heat in the exhaust steam is usually only about one-half of that in the impulse turbine—on the assumption of equal peripheral speeds—it follows that the reaction turbine must have about twice as many pressure stages as an equivalent impulse turbine.

The unlimited number of stages and the lower steam velocity are advantages which are of importance in reducing the number of revolutions. Velocity stages with reaction turbines are inapplicable.

### 3. Passage of the Steam in Turbines.

(a) In single-stage impulse turbines (Fig. 15), the passage of the steam through the vanes of the rotor must be effected in such a way that no alteration of pressure can take place, the wheel therefore revolving in steam under a very low pressure. Similarly the jet of steam must flow freely through the vanes as the steam works simply by impulse, so that the passages between the vanes must be of a fixed section, and the steam must enter so far as possible without shock and flow through the vanes without abrupt change of curvature.

The steam may be admitted to the rotor over the whole circumference (*full admission*), or only over a part of it (*partial admission*), and the direction of the jet of steam may form a fixed angle with the casing or with the face of the wheel.

(*b*) In multiple-stage impulse turbines the volume
of steam flowing per second is least in the first set of
nozzles, but increases more and more by expansion
in each stage. On account of this the section of the
first set of nozzles can be made of smaller dimensions and admit of partial admission to the first

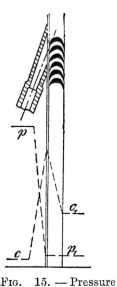

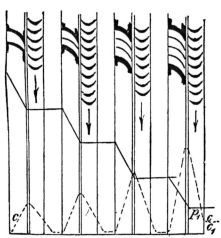

Fig. 15. — Pressure
and Velocity Changes
in the Single-stage
Impulse Turbines.

$pp_1$ = line of pressure
variation.
$cc_1$ = line of pressure
velocity.

Fig. 16. — Pressure and Velocity
Changes in a Multiple-stage Impulse Turbines (Zölly, Rateau).
(4 Pressure Stages with Increasing Admission.)

wheel. With the increasing volume of the steam
the admission must be extended to a greater portion
of the wheel circumference, in order finally to become
complete at the last wheel.

If the employment of further pressure stages is
necessary, either the vanes of the rotor must be
extended radially to provide a larger sectional area of

the passage, or if this is not permissible, on account of the excessive length of the vanes, a greater fall of pressure for conversion into kinetic energy must be arranged for in the following stages; this gives a greater steam velocity, and consequently a smaller passage is necessary. A larger diameter of rotor may also be chosen, this being usually accompanied by increased fall of pressure.

Larger areas are required for the outlet of a rotor than for the preceding guides, and the guides must be so shaped as to permit the change of direction of the

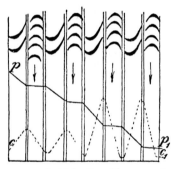

Fig. 17.—Pressure and Velocity Changes in a Full-admission Impulse Turbine (Hamilton-Holzwarth).

jets of steam coming from the rotor to take place with the greatest possible regularity over the breadth of the wheel, so that the distribution of kinetic energy increases uniformly. Each rotor runs in a chamber, which is shut off by diaphragms and in which the nozzles for the preceding and succeeding pressure stages are arranged. Only the last wheel runs in steam at the lowest pressure.

(c) In impulse turbines with velocity staging (Fig. 18) only a part of the motive power stored up in the steam is utilized in a single rotor (as stated on page

24), and therefore the steam issuing from the wheel still has sufficient velocity to be admitted, after the reversal of its direction in the guide wheel vanes, to a second and third wheel of the same pressure stage. The first pressure stage of such a turbine differs from a single-stage impulse turbine only in that the steam, on leaving the rotor, has not yet attained its lowest pressure but is capable of sustaining an additional fall of pressure. The nozzles of the last

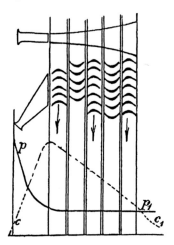

Fig. 18.—Pressure and Velocity Changes in a Multiple-stage Turbine with Velocity Stages (Curtis, A.E.G.).

pressure stage are enlarged to such an extent that the steam can expand to the pressure of the condenser. Against the advantage arising from the resulting decreased peripheral velocity, there is on the other hand the disadvantage that the losses through shock on entering the second and following rotors, as well as the losses set up by the reversal of the flow of steam in the guide wheel, are considerable.

(d) In reaction turbines (Fig. 19) the guide vanes

control the direction of flow of the entering steam for the purpose of causing the steam to impinge on the vanes of the rotor at the requisite angle. Expansion takes place in the guide and rotor vanes and the pressure in each rotor is lower than that in the preceding guide wheel. It follows therefore that the pressure and contained heat of the steam constantly diminishes, whereas the velocity increases as shown.

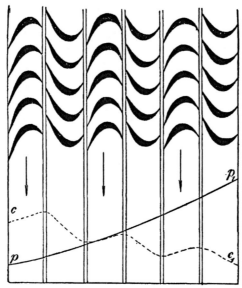

Fig. 19.—Pressure and Velocity Changes in a Reaction Turbine (Parsons).

The sectional area of the passages in each vane rim must therefore be increased on account of the growing volume of the steam; and this can be done either by altering the position of the vanes, thus modifying the inlet and outlet angle, and also the pitch of the vanes, or by suddenly enlarging the radial length of the vanes and the diameter of the wheel, with which arrangement can be combined an alteration of the pitch of the vanes. To avoid increasing the length of the passages, and therefore the friction, at the last stage, the outlet angle of the vanes is increased ; for, though small inlet and outlet angles of the guide and running vanes imply a diminished number of stages (because a larger fall of heat can be utilized in long passages), there would be considerably increased friction under these conditions.

As regards the number of stages, the reaction turbine is at a disadvantage as compared with the impulse turbine, because it cannot be run with partial admission, and therefore the first wheel must be made of small diameter, since otherwise the vanes would be too small to be practicable.

If the peripheral velocity of a reaction turbine and an impulse turbine be taken of equal magnitude, then, in the former case, the average value of the peripheral velocity of the individual wheels will be smaller, but the number of stages much larger than with the impulse turbine.

The advantage of a large number of stages is that, even with an unfavourable velocity ratio, the steam is enabled to give up all its available energy.

Figs. 15 to 19 give the pressure and velocity changes in different systems, the lines $pp_1$ representing the pressure and $cc_1$ the velocity.

The rotors are indicated by arrows. Besides the arrangements mentioned many other combinations are manufactured and patented; as, for example, combined impulse and reaction turbines, the former as high-pressure turbines (chiefly with velocity staging), the latter as low-pressure turbines. In others again the constricted portion of the nozzle is situated in the rotor and the widened portion in the following guide wheel, so that in the latter an increase in pressure takes place which is reconverted into velocity in the following rotor.

The Curtis and A.E.G. turbines are mostly made with only two pressure stages each of which has two velocity stages. In high powers there are, of course, more pressure stages on the low pressure side.

Velocity staging on the high pressure side affords a simple means of passing quickly from high boiler pressure and superheat to low pressure. The conversion of the superheat is attained in a very simple manner and the change of heat occurs in the nozzle.

### 4. Losses of Velocity, Pressure, and Steam in the Steam Turbine.

The losses of steam during its passage through the steam turbine are chiefly :—

Velocity losses due to steam friction in the nozzles and rotor vanes; losses by shocks, eddies, and vibration; clearance loss; loss of heat through the walls of the casing; loss through the friction of the rotors in the surrounding steam; losses on issuing from the last rotor; stuffing-box losses; losses through friction in the bearings and stuffing boxes; possible losses in gearing down; and losses from the power required to drive the condenser.

(a) **Losses in the nozzle.**—Even in well-constructed nozzles a reduction of the steam velocity is caused by the friction between the steam of the nozzle walls, which reduction militates against an exact adiabatic flow. Notwithstanding this, the pressure and velocity follow a uniform curve. The smallest section remains unaltered, owing to the low velocity existing there against the frictionless nozzle. Towards the end of the outflow, however, the section must be increased owing to friction in the nozzle, if the same amount of steam is to pass through, the velocities having increased.

The loss here, as in cylindrical pipes, can be taken as proportional to the length of nozzle and the square of the average velocity of flow, whilst inversely proportional to the average diameter of the nozzle, and amounts to about 10 to 15 per cent.

The loss is greatest in impulse turbines, where the whole fall of pressure is converted into kinetic energy.

(b) **The losses of velocity in the rotor vanes** are due to the same causes as in the nozzle ; and as in the case of the guide vanes will be so much smaller in proportion as the change of direction of the steam is more regular, and the conversion of the heat into work is more uniform.

(c) **Losses through shocks, eddies, and vibration.**—In partial admission especially the current of steam impinges also on portions of the guide wheels unprovided with vanes, and is partly dissipated, thus preventing uniform action of the steam in the guide and rotor vanes.   Hence, the greater the velocity the greater are these losses, i.e. in turbines with a small number of stages they will be greatest.

In reaction turbines the reaction effect entails compression of the current of steam against the side of the rotor blades, so that the steam density is greatest on the surface of the blades and less in the other and more remote portions of the section.   In

the succeeding guide wheel, the density ratio is
reversed, since the impact of the current of steam
acts against the opposite side of the vanes ; and as
this alternation is continuously repeated, shocks and
eddies are produced. Nevertheless, reaction turbines
may run more evenly than impulse turbines, the
opportunities for shocks and eddies being fewer ;
and this appears to demonstrate that the partial-admission impulse turbine does not possess any superi-
ority over the full-admission reaction turbine.

(*d*) **Clearance losses :—**

(*a*) In single-stage impulse turbines (De Laval,
Fig. 15), the steam which flows through the nozzle
but does not impinge upon the wheel possesses no
useful kinetic energy. A reduction of the wheel friction can be brought about by superheating steam
in the turbine casing, and by giving the laterally
escaping steam a rotary movement in the direction
of the rotor. The gain of power attained in this way
is, however, immaterial in comparison with the clearance loss.

In turbines with a single-pressure stage and
several velocity stages the clearance losses are
always fairly large on account of the increase in
the volume of steam flowing through.

In multi-stage turbines, the rotors of which run
in closed chambers under constant pressure, steam
losses through leakage of steam can result only from
the play between the shaft and the diaphragm
carrying the guide blades (Figs. 8, 9). Such play
can, however, be reduced to very small dimensions.

(*β*) In reaction turbines, steam losses occur in the
space between the rotor vanes and the casing and

between the shaft and the guide vanes, especially at the steam-inlet end.

The first of these losses could be avoided if the rotor were arranged for impulse working instead of for reaction; but the second loss would then be correspondingly the greater.

To minimize unintentional leakage of the steam from one stage to the other, there is, however, no necessity to reduce the radial play to its smallest possible dimensions, because energy in the form of heat is supplied to the working steam by the leakage steam, so that the former is superheated and consequently drier—a condition that appears desirable on account of the large proportion of moisture contained in the steam as the result of adiabatic expansion.

(e) **The losses from radiation of heat** are greater in such parts of the turbine where the steam temperature is high, than in those at lower temperatures, the hotter steam radiating a greater amount of heat than the cooler steam.

Accordingly, in systems with only a single row of expansion nozzles, the loss from heat radiation is less than in the remaining systems. In longer turbines the loss must evidently be greater than in shorter types.

(f) **Friction losses.**—The friction losses in flowing through the nozzles and vanes are difficult to ascertain; and no satisfactory data are at present available. These losses will be much greater for higher velocities, and they vary approximately as the square of the velocity. The losses will be less with a single nozzle than where two are used of a total area equal to the first one. Friction is also increased by using damp steam, or steam of higher pressure, as well as by the use of several rows of nozzles. With low velocities the friction may be taken as directly proportional to the velocity; but with higher velocities it is proportional to the square of same, and with very high velocities it increases in a higher ratio than the square of the velocity.

Turbines with two velocity stages for each pressure stage, however, sustain greater losses through steam friction at the vanes than those with only one velocity stage for each pressure stage, and their efficiency must therefore be lower. If in an impulse turbine the conversion into kinetic energy takes place in a single nozzle only, and the reduction of the peripheral velocity is attained by means of velocity staging, then as all the rotors work in exhaust steam the turbine runs under theoretically advantageous conditions. In reaction turbines all the vane spaces in the guide and rotor wheels are filled with steam, and the friction losses are small.

At the wheel periphery friction occurs due to the radial velocity of the steam caused by centrifugal force, although this perhaps is unimportant.

Of far more importance are the losses through the friction between the faces of the rotors and the surrounding steam. The losses chiefly arise from surface friction and eddies, which are difficult to determine separately with accuracy, but must be lessened by using smooth, unoxidizable discs. Trials with the De Laval turbines have shown that the work absorbed by resistance to rotation is almost exactly proportional to the density of the surrounding steam, and increases as the fifth power of the

wheel diameter and the cube of the number of revolutions. From this the friction losses vary as $d^5 n^3$
and the peripheral velocity varies as $dn$, so that for a
given peripheral wheel velocity, $d^3 n^3$ will be constant,
and the friction losses vary as $d^2$, i.e. the wheel surface, for which reason small wheels, run at a high
speed, are more favourable as regards friction losses
than larger wheels running more slowly. If a turbine, for example, with one pressure stage is compared with a similar one of $z$ pressure stages, the
number of revolutions of the latter is $\sqrt{z}$ times less
than the former. However, if the average density of
the steam in which the wheels of the multi-stage
turbine revolves be greater than in the single-stage turbine, the friction losses will also be greater. Another
factor which influences the question is the dryness
fraction $x$, for in using dry saturated steam only
the last wheel of the multi-stage impulse turbine will
be running in steam of the same dryness fraction as
in the single-stage impulse turbine. All other
wheels run in drier steam. Since, however, damp
steam increases the friction loss, it operates in favour
of the multi-stage impulse turbine, though nevertheless the influence of a higher steam density is always more important than a higher dryness fraction.
In reaction turbines with vanes arranged, as usual,
on the circumference of revolving drums (Fig. 13),
the face at the admission end of the turbine is protected from friction by the steam, so that the resistance to turning works out very small.

As the wheel vanes have a high peripheral velocity
they cause eddies in the surrounding steam similar
to those which occur with a fan. With a number

of vanes represented by $i$, the amount of work consumed in this fan or "windage" resistance may be expressed as approximately proportional to the product $\beta \rho i u^3$, in which $u$ = the peripheral velocity, $\rho$ = the density of the steam, and $\beta$ = a constant determined by experiment.

In impulse turbines a flow of steam probably is set up in each steam chamber, the result being a difference between the pressures at the centre and at the circumference, whilst an oppositely directed flow is set up along the face of the diaphragm towards the centre. The heat arising from this movement and from friction is only imperfectly utilized in the succeeding expansion, so that the output of the rotors is reduced.

With increasing superheat, under otherwise similar conditions, the loss by friction, though considerable, diminishes.

According to Stodola, the horse-power at no load (P′) (wheel friction losses, i.e. the sum of the fan or "windage" resistance and friction of the disc) for a disc of diameter D and a peripheral velocity $u$, both in foot-units, is approximately

$$ \mathrm{P}' \; = \; a \, . \, \mathrm{D}^2 \rho \left( \frac{u}{100} \right)^3 \qquad (21)^1 $$

[1] This formula is an approximate one for encased rotors.

The more exact formula given by Stodola for discs in air not encased is,

$$ \text{H.-p.} = \rho [0 \cdot 02295 \, \alpha_1 \, \mathrm{D}^{2 \cdot 5} + 1 \cdot 4246 \, \alpha_2 \, \mathrm{L}^{1 \cdot 25}] \left( \frac{u}{100} \right)^3 $$
$$ \alpha_1 = 3 \cdot 14 \text{ and } \alpha_2 = 0 \cdot 42 $$

L is the length of blades in inches.

In C. G. S. units $\rho$ in kg. per cm.$^3$, D in metres, L in cm. and $u$ in metres per second the same formula is used, but the co-efficients 0·02295 and 1·4246 are omitted.—[Ed.]

the coefficient $a$ being approximately determined on the assumption that the loss by friction, when the wheel is running in saturated steam, under otherwise equal conditions, is about 30 per cent greater than when revolving in air of the same density. For superheated steam the degree of superheating has, moreover, a considerable influence. For saturated steam $a$ may be taken as 0·042, $\rho$ being in lb. per cubic foot.

(g) **Centrifugal force on water.**—The water set in motion by the action of the centrifugal force may cause a loss of energy in steam turbines, especially when saturated steam is used, so that efficient drainage of the working chamber of each stage is necessary. This can be done by means of tangential openings in the direction of the moving water. The smaller the number of stages and the area of the clearances between casing rotors, and the higher the peripheral and steam velocities used, the more important does an efficient drainage become.

(h) **Exhaust losses.**—The steam on leaving the turbine still possesses a certain velocity U ; and the amount of useful work passing from the turbine, on this account, amounts to

$$\frac{U^2}{2g} \text{ foot-lb. per 1 lb. of steam.}$$

The total work losses in the nozzles, the rotor vanes, and at the exhaust are converted into heat, so that the dryness fraction of the steam increases. If V and U represent the inlet and outlet velocities at the nozzles $V_1$, $V_2$ etc., $U_1$, $U_2$ etc., representing the velocities on entering and leaving the vanes of successive rotors, and U = the steam velocity on issuing from the turbine, then the loss of heat energy during the passage of 1 lb. of steam is equal to

$$E = \frac{V_0{}^2 - U_0}{2g} + \frac{V_1{}^2 - U^2}{2g} + \frac{V_2{}^2 - U_2{}^2}{2g} + \ldots + \frac{U^2}{2g}$$

$$(22)$$

In any stage if $x$ and $x_1$ are the dryness fractions of the steam, $t$ and $t_1$ the corresponding temperatures, and L and $L_1$ the corresponding latent heats, then the heat drop in work units for the stage is given by

$$E = t + xL - t_1 - x_1L_1 \qquad (23)$$

By a combined use of equations 22 and 23 we can get a relation between the change of velocity and of the other properties of the steam in any stage.

## 5. GRAPHICAL DETERMINATION OF THE INLET AND OUTLET VELOCITY.

In Figs. 20 to 24 the velocity diagrams of the separate types of turbines are shown in combination with the corresponding entropy diagrams.

The absolute intake velocities are shown by V, $V_1$, $V_2$; the peripheral velocities by $u$, $u_1$, $u_2$; the absolute outlet velocities by U, $U_1$, $U_2$; the relative inlet and outlet velocities by R, $R_1$, $R_2$ and $r$, $r_1$, $r_2$; the inlet angle of the vane by $a$, $a_1$, $a_2$, and the outlet angle by $\beta$, $\beta_1$, $\beta_2$.

The angles relate to the side of the vanes upon which the steam impinges. Friction and other losses are left out of consideration.

The amount of heat converted into kinetic energy and supplied to the separate rotors is proportional to the squares of the inlet velocities, i.e. $V^2$, etc., whilst the amount rejected is proportional to $U^2$.

In an ideal turbine the outlet velocity should be zero, but this is never achieved in practice.

The kinetic energy proportional to $U^2$ and still existing at the exhaust is again converted into heat, and must be added to the amount of heat, which cannot be converted into useful work, even in the ideal turbine.

## (*a*) Impulse Turbine with one Pressure Stage (Fig. 20).

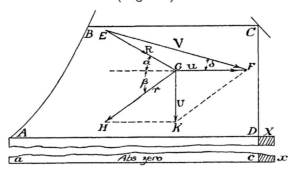

Fig. 20.—Entropy and Velocity Diagram for a Single-stage Impulse Turbine.

Amount of heat supplied to the nozzles . . Area $a$ABC$c$
Heat converted into energy in the nozzles . ,, ABCD
Heat unconverted into energy in the nozzles . ,. $a$AD$c$
Absolute intake velocity at the rotor . . . ,, V = EF
Peripheral velocity of the rotor . . . . ,, $u$ = FG
Relative intake velocity at the rotor . . . ,, R = EG
Relative outlet velocity from the rotor . . ,, $r$ = HG
Absolute outlet velocity from the rotor . . ,, U = KG

$$\text{Efficiency}: \eta = \frac{V^2 - U^2}{V^2}$$

## (*b*) Impulse Turbines with Three Pressure Stages (Fig. 21).

Amount of heat entering first set of nozzles . Area $a$ABC$c$
Heat converted into kinetic energy in the first set
  of nozzles . . . . . . . ,, LCBM

Kinetic energy converted back into heat at outlet from first rotor . . . . . . Area $c$MX$x$

Heat supplied to second set of nozzles . . ,, $a$ALX$x$

Heat converted into kinetic energy in second set
of nozzles . . . . . . . ,, NLXV

Kinetic energy converted back into heat at outlet from second rotor . . . . . ,, $x$VY$y$

Heat supplied to third set of nozzles . . . ,, $a$ANY$y$

Heat converted into kinetic energy at third set
of nozzles . . . . . . . ,, ANYT

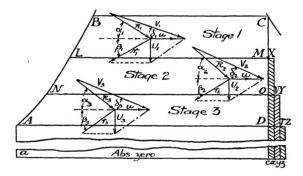

Fig. 21.—Entropy and Velocity Diagram for an Impulse
Turbine with Three Pressure Stages.

Kinetic energy converted back into heat at outlet from third rotor . . . . . . Area $y$TZ$z$

Total heat passing to condenser . . . . ,, $a$AZ$z$

Absolute steam velocity on leaving first set of
nozzles . . . . . . . . ,, $V_1$

Peripheral velocity of first rotor . . . . ,, $u$

Relative inlet velocity at the first rotor . . ,, $R_1$

Relative outlet velocity from the first rotor . . ,, $r_1$

Absolute steam velocity on leaving the second set
of nozzles . . . . . . . . ,, $V_2$

Absolute steam velocity on leaving the second
rotor . . . . . . . . ,, $U_2$

Peripheral velocity of the second rotor . . ,, $u$

Relative inlet velocity at the second rotor . . ,, $R_2$

Relative outlet velocity at the second rotor . ,, $r_2$

Absolute steam velocity on leaving the third set of nozzles . . . . . . . . Area V$_3$

Absolute steam velocity on leaving the third rotor ,, U$_3$

Peripheral velocity of the third rotor . . . ,, $u$

Relative inlet velocity at the third rotor . . ,, R$_3$

Relative outlet velocity at the third rotor . . ,, $r_3$

$$\text{Efficiency}: \eta = \frac{V_1{}^2 - U_1{}^2 + V_2{}^2 - U_2{}^2 + V_3{}^2 - U_3{}^2}{V_1{}^2 + V_2{}^2 + V_3{}^2}$$

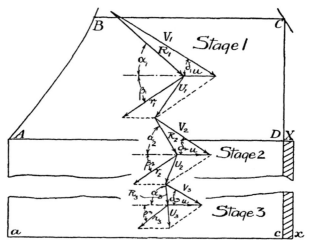

Fig. 22.—Entropy and Velocity Diagram for an Impulse Turbine with Three Velocity Stages.

## (c) Impulse Turbine with Three Velocity Stages (Fig. 22).

Steam expanding in the nozzles to the pressure of the condenser.

Amount of heat entering the nozzles . . . Area $a$ABC$c$

Heat converted into kinetic energy . . . ,, ABCD

Kinetic energy converted into heat at outlet . ,, $c$DX$x$

Total heat carried to condenser . . . ,, $a$AX$x$

Velocity of outflow from the nozzles . . . ,, V$_1$

Peripheral velocity of the first rotor . . . ,, $u$

Relative inlet velocity at the first rotor . . ,, R$_1$

Relative outlet velocity at the first rotor . . ,, $r_1$

Absolute outlet velocity at the first rotor  .    .  Area    $U_1$

Reversal of the direction of flow in the guide
vanes.

Inflow velocity at the second rotor   .    .    .     ,,  $V_2 = U_1$

Outflow velocity at the second rotor .    .    .     ,,      $U_2$

Reversal of the direction of flow in the guide
vanes.

Inflow velocity at the third rotor    .    .    .     ,,  $V_3 = U_2$

Outflow velocity at the third rotor   .    .    .     ,,      $U_3$

$$\text{Efficiency}: \frac{V_1{}^2 - U_3{}^2}{V_1{}^2}$$

## (d) Impulse Turbine with Two Pressure Stages and Two Velocity Stages to Each (Fig. 23).

Total heat entering the first set of nozzles   .   Area $a$ABC$c$

Heat converted into kinetic energy in the first set
of nozzles  .    .    .    .    .    .    .     ,,   QBCD

Heat not converted into kinetic energy in the
first set of nozzles '   .    .    .    .    .     ,,   $a$AQD$c$

Kinetic energy converted into heat at outlet from
second rotor .    .    .    .    .    .    .     ,,   $c$DX$x$

Amount of heat entering the second set of
nozzles .    .    .    .    .    .    .    .     ,,   $a$AQX$x$

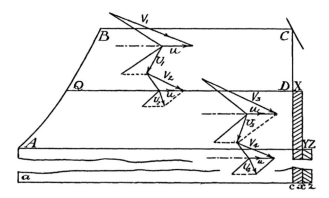

Fig. 23.—Entropy and Velocity Diagram for an Impulse
Turbine with Two Pressure Stages (Two Velocity Stages to
each).

Heat converted into kinetic energy in the
nozzles . . . . . . . . Area  AQXY

Heat carried away to condenser . . . „  $a$AY$x$

Kinetic energy converted into heat on leaving
the fourth rotor . . . . . . „  $x$YZ$z$

Total heat carried to condenser . . . „  $a$AZ$z$

Absolute inlet velocity at the first rotor . . „  $V_1$

Absolute outlet velocity at the first rotor . . „  $U_1$

Reversal of the direction of flow in the guide vanes.

Absolute inlet velocity at the second rotor . „  $V_2 = U_1$

Absolute outlet velocity at the second rotor . „  $U_2$

Absolute inlet velocity at the third rotor . „  $V_3$

Absolute outlet velocity at the third rotor . . „  $U_3$

Reversal of the direction of flow in the guide
vanes.

Absolute inlet velocity at the fourth rotor . „  $V_4 = U_3$

Absolute outlet velocity at the ourth rotor . „  $U_4$

$$\text{Efficiency:} \quad \eta = \frac{V_1^2 - U_2^2 + V_3^2 - U_4^2}{V_1^2 + V_3^2}$$

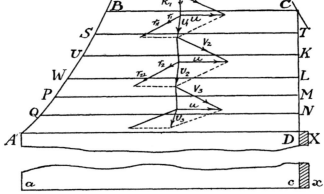

FIG. 24.—Entropy and Velocity Diagram for a Multi-stage
Reaction Turbine.

## (e) Multi=stage Reaction Turbine (Fig. 24).

The steam expands gradually in the guide and
rotor wheels to the pressure of the condenser.

Amount of heat entering the first guide wheel . Area $a$ABC$c$

Heat converted into kinetic energy in the first guide wheel . . . . . . . ,, SBCT

Heat converted into kinetic energy in the first rotor . . . . . . . . ,, USTK

Heat converted into kinetic energy in the second guide wheel . . . . . . . ,, WUKL

Heat converted into kinetic energy in the second rotor . . . . . . . . ,, PWLM

Heat converted into kinetic energy in the third guide wheel . . . . . . . ,, OPMN

Heat converted into kinetic energy in the third rotor . . . . . . . . ,, AQND

Kinetic energy converted back to heat on leaving the third rotor . . . . . . ,, $c$DX$x$

Heat carried to condenser . . . . ,, $a$AX$x$

Outlet velocity from first guide wheel . . ,, $U_0 = V_1$

Peripheral velocity of the first rotor . . . ,, $u$

Relative inlet velocity in the first rotor . . ,, $R_1$

Relative outlet velocity from first rotor . . ,, $r_1$

Increased relative outlet velocity from the first rotor . . . . . . . . ,, $r_{11}$

Absolute outlet velocity from the first rotor . ,, $U_{11}$

Increase of velocity in the second guide wheel . ,, $V_2$

Relative outlet velocity from the second rotor . ,, $r_2$

Increased relative velocity from the second rotor ,, $r_{22}$

Absolute outlet velocity from the second rotor . ,, $U_2$

Absolute outlet velocity from third rotor . . ,, $U_3$

Efficiency :

$$\eta = \frac{V_1{}^2 + r_{11}{}^2 - r_1{}^2 - U_1{}^2 + V_2{}^2 + r_{22}{}^2 - r_2{}^2 - U_2{}^2 + V_3{}^2 + r_{33}{}^2 - r_3{}^2 - U^2{}_3}{V^2 + r_{11}{}^2 - r_1{}^2 - U_1{}^2 + V_2{}^2 + r_{22}{}^2 - r_2{}^2 - U_2{}^2 + V_3{}^2 + r_{33}{}^2 - r_3{}^2}$$

## 6. PERIPHERAL AND STEAM VELOCITIES OF THE DIFFERENT TYPES OF TURBINES.

(a) *Impulse Turbines.*—If a complete reversal of the direction of the steam velocity were effected in an ideal single-stage impulse turbine, then to get maximum efficiency the peripheral velocity $u$ must equal $\frac{1}{2}$ V; but in actual experience $u$ is always less.

V is frequently about 3900 feet per second, so that high velocities of the wheel are necessary.

In multi-stage impulse turbines with a velocity stage to each, the ideal condition would be $u = \frac{1}{2} V_1$, $u_2 = \frac{1}{2} V_2$, and so on; this, however, is not met with in practice.

Whether the ratio between $v$ and $u$ is in practice greater or less than 0·5, it is, however, approximately constant for all the stages of the same turbine.

For a multi-stage impulse turbine with four rotors the peripheral velocity is $\sqrt{4} = 2$ (and for a similar one with twenty-five rotors $\sqrt{25} = 5$) times less than that of a single-stage impulse turbine under the same conditions.

In an ideal single-stage impulse turbine with two velocity stages, the steam at each stage should have a velocity equal to double the peripheral velocity of the stage in question, so that, for example, in the first rotor, $V_1$ is reduced to $2u_1$ and the resulting steam velocity $U_1$ or $V_2$ is diminished to $2u_2$ in the second rotor.

If all the rotors had the same peripheral velocity, the most favourable initial velocity would be $\dfrac{V}{i}$, assuming the number of rotors to be $i$; the velocity must therefore be inversely proportional to twice the number of rotors.

Although this does not exactly fit in with practice, nevertheless a turbine with one pressure stage and several velocity stages must be considerably shorter than a similar one of equal peripheral velocity, but with several pressure stages and a velocity stage to each.

For turbines with several pressure stages and several velocity stages to each, the peripheral velocity for $m$ pressure stages and $q$ velocity stages to each is inversely proportional to

$$2 \, q \, \sqrt{m}$$

Bearing friction in mind, the following condition should be satisfied in impulse turbines, viz. that the sectional area of the passage through the vanes must be equal to or greater than the product of the inlet areas and the ratio of the inlet velocity to the outlet velocity, allowing for friction.

(b) *Reaction Turbines.*—In reaction turbines with an equal heat drop for each stage (guide and rotor) approximately the same maximum steam velocity (taking the peripheral velocity of the rotor as constant) in passing through the guides and rotors can be attained throughout.

For given steam velocities $V_1$, $V_2$, etc., the rotor velocity will be twice as great as in impulse turbines with several pressure stages (each with a velocity stage). Nevertheless, given an equal number of stages, the steam velocities in the latter are twice as great, so that in this case reaction turbines do not possess much higher peripheral velocities, these—neglecting losses due to friction and clearance—being inversely proportional to the square root of the number of rotors.

### 7. Output, Efficiency, Steam Consumption.

### (a) Output and Efficiency.

*Impulse Turbines.*

The theoretical work done in a turbine is determined by the available loss of heat.

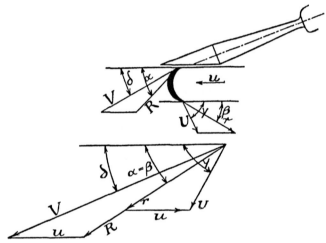

Fig. 25.—Velocity Diagram for One Pressure Stage and One Velocity Stage.

If $V_o$ = the theoretical value of one lb. of steam flowing through the nozzle, and $V$ = the actual inlet velocity at the rotor, then the work supplied to the turbine has the value

$$E_o = \frac{V^2}{2g}\left(\frac{1}{V^2 \div V_o^2}\right) \qquad (24)$$

or expressed from the loss of heat

$$E_o = 778 \ (H_o - H_1) \ \text{foot-lb.}$$

According to Mollier the available heat drop in frictionless adiabatic expansion from $p_o$ to $p_1$, is

$$= H_0 - H_1 = 2544 \frac{\log p_0 - \log p_1}{17\cdot 68 - 2\cdot 011 \log p_1} \text{ B.Th.U.}$$

per lb. steam          (26)

(a) *Turbine with One Pressure Stage and One Velocity Stage (De Laval) (Fig. 25).*

The work per lb. steam, given out at the wheel rim according to the velocity diagram, when V and U represent the absolute inlet and outlet velocities, and $\delta$ and $\gamma$ their angles of inclination (Fig. 25), amounts to

$$E_i = \frac{1}{g} (V \cos \delta + U \cos \gamma) u \qquad (27)$$

This can be simplified to

$$E_i = \frac{Vu}{g} \left( \cos \delta - \frac{u}{V} \right) \left( 1 + \frac{r}{R} \right) \qquad (27a) [1]$$

therefore we get as velocity or "indicated" efficiency

$$\eta_i = \frac{E_i}{E_0} \qquad (28)$$

or putting in our value of $E_0$ from equation (24)

[1] This can be seen by remembering that the lower diagram of Fig. 25 is a combined velocity diagram, the outlet triangle of velocities being turned round.

Equation (27) is the ordinary turbine formula based upon the change of angular momentum.

To bring equation (27a) to that in equation (27) imagine the line $u$ of the smaller triangle to be produced to meet the line V.

Then

$$27a = \frac{u}{g} \left( V \cos \delta - u \right) \left( 1 + \frac{r}{R} \right) = \frac{u}{g} \left\{ V \cos \delta + \frac{V \cos \delta r}{R} - u - \frac{ur}{R} \right\}$$

Considering the projections it will be clear that

$$\frac{V \cos \delta r}{R} - u - \frac{ur}{R} = U \cos \gamma \therefore (27a) = \frac{u}{g} \left( V \cos \delta + U \cos \gamma \right)$$

$$\eta_i = 2 \left(\frac{V}{V_o}\right)^2 \left(1 + \frac{r}{R}\right) \frac{u}{V} \left(\cos \delta - \frac{u}{V}\right) \qquad (29)$$

If work absorbed by friction be neglected, $R = r$ and $V = V_o$, then

$$\eta_i = 4 \frac{u}{V}\left(\cos \delta - \frac{u}{V}\right) \quad \cdots \quad (30)$$

The maximum value of $\eta_i$ is found from equation (29) to occur when $u = \frac{1}{2} V \cos \delta$ and has the form

$$\eta_{i(\text{max.})} = \frac{1}{2}\left(\frac{V}{V_o}\right)^2\left(1 + \frac{r}{R}\right) \cos^2 \delta \qquad (31)$$

If $P_e$ be taken to represent the effective horse-power that the turbine should theoretically produce, and the work with no load be expressed by

$$P^1 = kP_e \qquad (32)$$

in which $k = 0 \cdot 2$ to $0 \cdot 3$ (approx.), the mechanical efficiency $\eta_m$ will be

$$\eta_m = \frac{P_e}{(1 + k) P_e} \qquad (33)$$

so that we get as overall or nett efficiency,

$$\eta_e = \eta_i \cdot \eta_m$$

The equations (29 to 41) for $\eta$, when plotted with $\eta_i$ as ordinates and $\frac{u}{v}$ as abscissæ, furnish a curve similar to a parabola, whose highest point gives the maximum value of $\eta_i$.

### (β) Turbine with One Pressure Stage and Two or more Velocity Stages (Fig. 26).

If the velocities corresponding to the first and second velocity stage be marked with the subscripts 1 and 2 respectively, then the sum of the efficiencies of the separate rims represents the efficiency of the turbine, thus:

$$\eta_i = \frac{E_{i_1} + E_{i_2}}{E_0} \qquad (35)$$

or
$$\eta i = \eta i_1 + \eta i_2 \qquad (36)$$

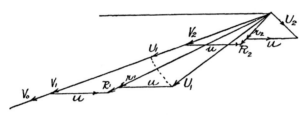

FIG. 26.—Velocity Diagram for One Pressure Stage and Two Velocity Stages.

In the above, $\eta_1$ can be determined from equation (29), whilst for $\eta_{i_2}$ we have

$$\eta_{i^2} = 2\left(\frac{V_2}{V_0}\right)^2\left(1 + \frac{V_2}{U_1}\right)\frac{u}{V_2}\left(\cos \delta_1 - \frac{u}{V_2}\right) \qquad (37)$$

The efficiency attainable diminishes as the number of stages is increased.

($\gamma$) *Turbines with Several Pressure Stages (Rateau, Zölly).*

With a pressure $p_1, p_2, p_3, \ldots$ in the separate stages, the available fall in each amounts to

$$p_0 - p_1, \ p_1 - p_2, \ p_2 - p_3, \text{ etc.}$$

For $m$ stages the heat drop in each stage is

$$\frac{E_o}{778\ m}$$

and the values of $\eta_i$ corresponding to this fall must be determined.

The corresponding mechanical efficiency is obtained from equation (33).

If $\gamma_1$, $\gamma_2$, $\gamma_3$, etc., represents the density of the steam in the separate stages, then, according to Stodola,

$$E_o = a \cdot D^2 \left(\frac{u}{100}\right)^3 (\gamma_1 + \gamma_2 + \gamma_3 + \gamma_m) \qquad (38)$$

and therefore again,

$$\eta_e = \eta_i \cdot \eta_m$$

For the best nett efficiency the peripheral velocity for the number of stages to be chosen is the determining factor.

### ($\delta$) Turbines with Several Pressure and Velocity Stages (Curtis).

The adoption of a second velocity stage reduces the number of the pressure stages but increases the indicated efficiency, provided $\frac{u}{V} < \cdot 273$.

### Reaction Turbines (Fig. 27).

The number of pressure stages being large, the inlet velocity in a guide channel can be set down as equal to the absolute outlet velocity of the rotor in the same stage.

If the entering angle of the absolute inlet velocity $V$ into the rotor and the outlet angle of the relative outlet velocity $r$ from the rotor are equalized, and

if the components of the velocities in the direction
of the axis be assumed as equal in each stage, in
consequence of the constant blade lengths and

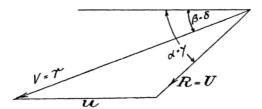

FIG. 27.—Velocity Diagram for a Reaction Turbine.

strength, then, with equal coefficients of resistance,
we can get for $E_o$ :

$$E_o = \frac{1}{g} (\xi V^2 - u^2 + 2u \; V \cos \delta) \text{ in foot-lb. per lb.}$$

steam. $\xi$ is a coefficient representing the work lost
in overcoming resistances $\xi$ and can be taken as $0 \cdot 25$
to $0 \cdot 45$.                                    (39)

The indicated work $E_i$, transmitted to the rotor per
lb. steam, is then

$$E_i = \frac{u}{g} (V \cos \delta + U \cos \gamma) = \frac{u}{g} (2 \; V \cos \delta - u)$$
(40)

consequently the indicated efficiency at each stage

$$\eta_i = \frac{2 \dfrac{u}{V} \cos \delta - \left(\dfrac{u}{V}\right)^2}{\xi + 2 \left(\dfrac{u}{V}\right) \cos \delta - \left(\dfrac{u}{V}\right)^2}$$      (41)

$$\text{If } \frac{u}{V} = \cos \delta$$

$n_i$ has a maximum value.

If $U_z$ represent the outlet velocity from the last
rotor, then

$$E_i = 778 \,(1 - \xi)\,(H_o - H_z) - \frac{U^2_z}{2g} \text{ foot-lb.} \qquad (42)$$

or

$$P = 1{\cdot}42 \,W \,(1 - \xi)\,(H_o - H_z) - \frac{WU^2}{35200} \text{ H.-P.} \qquad (43)$$

where W = weight of steam passing per second.

### Practical Determination of the Effective Output (B.H.-P.).

The indicated output of a turbine cannot be directly ascertained by testing, and can only be determined by estimation from the effective output.

In comparing steam turbines and reciprocating engines the effective output of the latter must be taken as the basis; but this is seldom done, owing to the difficulty in obtaining reliable data. The effective thermodynamic efficiency forms the only exact means of comparison, that is to say, the ratio between the work given out at the engine shaft per 1 lb. of steam, and the theoretical work that the steam is capable of doing in its expansion.

In steam turbines for driving dynamos the B.H.-P. can be determined very conveniently, because this can be easily arrived at from the output per effective kilowatt-hour, and is determined by direct reading from the electrical measuring instruments, the efficiency of the dynamo being known. This B.H.-P. can then be compared with the feed water consumption for an overall efficiency.

If $\eta_e$ and $\eta_d$ represent the efficiency of the steam turbine and dynamo, then 1 kw. of the dynamo corresponds to an output of

$$\frac{1000}{746 \ \eta e \eta_d}. \ \text{H.-P.}$$

of the turbine.

The direct measurement of the effective power of a steam turbine is made with a torsion dynamometer, the readings of which depend upon the relative angular twist of a measured length of the shaft. Such dynamometers have been described by Föttinger, Denny, and others.

### (b) Steam Consumption.

(a) For determining the theoretical steam consumption in lb. per hour, $C_0$ for 1 H.-P., the following formulæ have been given by Mollier and Rateau for saturated steam :—

According to Rateau,

$$C_0 = 1 \cdot 9 + \frac{17 \cdot 91 - 2 \cdot 056 \ p_0}{\log p_0 - \log p_1} \qquad (44)$$

According to Mollier,

$$C_0 = \frac{17 \cdot 68 - 2 \cdot 011 \log p_2}{\log p_0 - \log p_1} \qquad (45)$$

This theoretical consumption of steam should be compared with the actual measured consumption in connection with the electrical H.-P., in the case of driving dynamos, or with the work of lifting water in the case of pumps. When superheated steam is used, the heat energy corresponding to the superheating must be taken into consideration. In this connection it may be noted that recent experiments indicate that the specific heat of steam is not constant but diminishes with the amount of superheat. For low superheats its value is probably about unity, whilst for higher superheats it approaches 0·49.

(b) In all types of turbines the effective total consumption of steam, $C_e$, varies almost uniformly with the alteration of the load, and in this respect impulse and reaction turbines differ but little from each other. According to published experimental data, the reaction turbine is generally superior to the impulse turbine, because the process of expansion in the former is gradual, whilst in impulse turbines the possibility of losses through eddies is greater.

The indicated steam consumption $C_i$ may be set down for impulse turbines at

$$C_i = \frac{2544\cdot 6}{H_o - H_1} \text{ lb.} \qquad (46)$$

per theoretical or " indicated " H.-P. hour, and for reaction turbines as

$$C_i = \frac{1980000}{778\ (H_o - H_1) - \dfrac{U_z^2}{2g}} \qquad (47)$$

per " indicated " H.-P. hour.

The effective steam consumption for B.H.-P. hour is therefore

$$C_e = \frac{1}{\eta_e} \cdot C_i \text{ lb.} \qquad (48)$$

$C_e$ diminishes with increasing steam pressure, increased superheating and lower back pressure.

Superheating increases the volume of the steam, thereby diminishing the leakage losses. The steam consumption is about 1 per cent less for 9 to $12\frac{1}{2}°$ F. of superheating; but an expenditure of about 0·5 per cent more heat is necessary.

If the fall of heat is to remain constant for all outputs of a steam turbine, and the condition of the steam therefore remains the same, then, if any modification of the output be desired, the amount of steam flowing through the turbine must be changed.

For the maximum output, $E_o$ and W are fixed values, and therefore the sectional areas of the passages are also fixed, so that for a definite steam ratio only a well-defined conversion of energy can take place.

### TABLE IV.

THE RELATIVE THERMODYNAMIC EFFICIENCIES OF VARIOUS TYPES OF TURBINES UNDER DIFFERENT LOADS, 100 BEING TAKEN AS THE EFFICIENCY AT FULL LOAD.

| | Full Load in kw. | Full Load. | $\frac{3}{4}$ Load. | $\frac{1}{2}$ Load. | $\frac{1}{4}$ Load. |
|---|---|---|---|---|---|
| 1. Zölly . . . . | 300 | 100 | 80·3 | 58·8 | 36·3 |
| 2. Rateau . . . | 1000 | ,, | 78·0 | 55·6 | 33·7 |
| 3. ,, . . . | 500 | ,, | 77·2 | 55·0 | 32·0 |
| 4. Parsons-Westinghouse | 1250 | ,, | 78·5 | 57·3 | 36 |
| 5. ,, ,, | 400 | ,, | 78·5 | 57 | — |
| 6. Parsons . . . | 1500 | ,, | 78·0 | 56 | 33·4 |
| 7. ,, . . . | 500 | ,, | 78·0 | 56 | 33·7 |
| 8. Brown-Boveri-Parsons | 3000 | ,, | 77·6 | 56 | 33·3 |
| 9. Curtis . . . | 2000 | ,, | 77·2 | 54·2 | 31·4 |
| 10. ,, . . . | 600 | ,, | 76·5 | 52·4 | 28·8 |
| 11. ,, . . . | 500 | ,, | 76·2 | 51·7 | 28·1 |
| 12. Laval . . . | 200 | ,, | 76·2 | 52·3 | 29 |
| 13. ,, . . . | 20 | ,, | 82 | 68 | 52 |

Table V gives results of the trials for determining the steam consumption of a 600 H.-P. (400 kw.) Parsons-Westinghouse turbine.

## TABLE V.

### Results of the Trials for Determining the Steam Consumption of a 600 H.-P. (400 Kw.) Parsons-Westinghouse Turbine.

| | 2 | 1·5 | 1·25 | 1 | 0·75 | 0·5 | |
|---|---|---|---|---|---|---|---|
| 1. Loads | | | | | | | |
| 2. Steam pressure lb. per sq. in. (atmos.) | 157·0 155·8 | 154·5 156·4 | 158·3 156·4 | 157·7 157·7 | 158·3 156·? | 158·3 159·2 | superheated steam saturated   ,, |
| 3. Superheating °F. | 100 2·4 | 100·3 0·76 | 92 2·9 | 100·3 1·8 | 99 1·8 | 92·6 2·9 | superheated steam saturated   ,, |
| 4. No. of revolutions | 3454·5 3496·1 | 3460·8 3500·3 | 3486 3513·3 | 3502·8 3571·3 | 3532·2 — | 3561·6 — | superheated steam saturated   ,, |
| 5. Steam consumption lb. per h.-p. per hour | 13·5 15·1 | 12·7 — | 12·4 13·8 | 12·43 13·8 | 13·4 14·9 | 14·3 15·8 | superheated steam saturated   ,, |
| 6. Efficiency per cent | 53·7 51 | 56·4 — | 58·7 55·7 | 58·5 55·5 | 54·3 51·2 | 51·3 48·6 | superheated steam saturated   ,, |

8. TYPICAL CALCULATION FOR A STEAM TURBINE
   WITH THREE VELOCITY STAGES FOR AN OUTPUT
   OF 50 B.H.-P.

Given : Admission pressure $p_0$ = 128 lb. per sq. in.
        Condenser pressure $p_1$ = 1·4 lb. per sq. in.
        Revolutions per minute $n$ = 3000
Peripheral velocity of the rotor $u$ = 380 feet per sec.

Four nozzles are to be used, and the velocity of the steam is to be utilized by repeated admission in each wheel (Fig. 11). As a basis the steam consumption per hour may be taken as 18·7 lb. with a 10 per cent loss at each nozzle, for each reversal and each passage through the rotor vanes.

(*a*) The diameter D of the rotor can be calculated

from
$$\frac{D \pi n}{60} = u$$

$$D = \frac{60 \times 380}{\pi \cdot n} = 2\cdot41 \text{ feet.}$$

(*b*) The available energy per 1 lb. of steam, with adiabatic expansion, amounts to 224516 ft.-lb. corresponding to 288·6 B.Th.U.

(*c*) The outflow velocity from the nozzle amounts to 3775 feet per second (as per Table II), or, allowing for 10 per cent loss, approximately 3400 feet.

With a nozzle angle of 16° and the angle of 20° for the vanes, we have (allowing for loss in each case) :—

5

| | Relative Inlet Velocity. Feet. | Relative Outlet Velocity. Feet. | Loss of Energy due to Flow. Ft.-lb. per lb. of Steam. |
|---|---|---|---|
| Nozzle | | 3400 | $(3775^2 - 3400^2) \div 2g = 42312$ |
| Rotor I | $0 \cdot 9 \times 3050 = 2745$ | 1948 | $(3050^2 - 2745^2) \div 2g = 27552$ |
| Reversing Vanes | $0 \cdot 9 \times 2165 = 1948$ | 1624 | $(2394^2 - 1948^2) \div 2g = 16236$ |
| Rotor II | $0 \cdot 9 \times 1804 = 1624$ | 1166 | $(1804^2 - 1624^2) \div 2g = 9250$ |
| Reversing Vanes | $0 \cdot 9 \times 1295 = 1166$ | 773 | $(1295^2 - 1166^2) \div 2g = 5018$ |
| Rotor III | $0 \cdot 9 \times 859 = 773$ | 548 | $(1166^2 - 773^2) \div 2g = 2624$ |
| Exhaust | | | $548^2 \div 2g = 4690$ |

$$\text{Total } E_L = 107682$$
$$\text{say} = 107700$$
$$= 138 \cdot 4 \text{ B.Th.U. per lb.}$$

(d) The dryness fraction of steam $x_o'$ at the end of the adiabatic expansion in the nozzle, provided dry saturated steam is admitted, amounts to $0 \cdot 772$, at a pressure $p_c = 1 \cdot 4$ lb. per sq. inch, the latent heat being $L = 1037$ B.Th.U. As the loss of energy is converted into heat the dryness fraction rises to $x$, which can be determined from equation (23) :

$$\frac{E_L}{778} = L(x - x_o)$$

$$x = \frac{E_L}{778L} + x_0 = \frac{107682}{778 \times 1037} + 0{\cdot}772 = 0{\cdot}905.$$

At a pressure of $P_c = 1{\cdot}4$ lb. per sq. feet a volume of 172 cubic feet per lb. corresponds to a steam density $p = 0{\cdot}0058$. The work when running at no load is obtained from equation (21):

$$E^1 = 0{\cdot}170 \times (2{\cdot}41)^2 \left(\frac{380}{100}\right)^3 \times 0{\cdot}0058 \ . \ \text{H.-P.}$$

$$= 0{\cdot}063 \ \text{H.-P.}$$

(*e*) As the total steam consumption per second amounts to

$$\frac{50 \times 18{\cdot}7}{3600} = 0{\cdot}26 \ \text{lb.,}$$

then the theoretical power at the shaft is

$$\frac{0{\cdot}26 \, (778 \times 288{\cdot}6 - 107700)}{550} - 0{\cdot}063 = 55 \ \text{H.-P.}$$ ap-

proximately, allowing about 5 per cent for other losses such as friction of bearings, etc., then the B.H.-P. = approximately 52.

(*f*) The dimensions of the nozzles are determined from equation (14) according to the amount of steam flowing through each one per second, $W \div 4 = 0{\cdot}065$ lb., and $p_0 : p_c = 90$ (according to Table III)

$$A_x = \frac{W_x}{42{\cdot}9}\sqrt{\frac{v_0}{p_0}} = \frac{0{\cdot}065}{42{\cdot}9}\sqrt{\frac{3{\cdot}45}{128}} = 0{\cdot}00249 \ \text{sq. feet.}$$

$$F_x = 0{\cdot}035 \ \text{sq. inch.} \qquad d_x = 0{\cdot}21 \ \text{inch.}$$

From Table III $\frac{d}{d_x} = 3{\cdot}72$, we get

$$d = 0{\cdot}78 \ \text{inch.}$$

For an internal angle of 15° we get, for the expansion length $l$ of the nozzle,

$$l = \frac{0\cdot39 - 0\cdot105}{0\cdot132} = 2\cdot16 \text{ inches.}$$

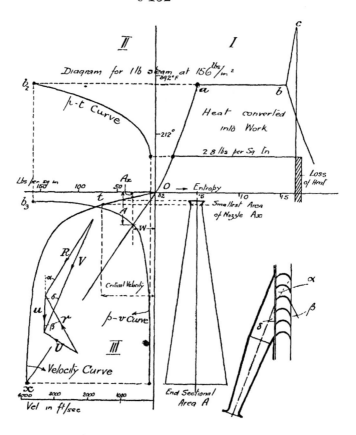

FIG. 28.—Graphical Determination of the Dimensions of Nozzles and the Steam Velocities of a Single-stage Impulse Turbine (De Laval).

## 9. Example of Graphical Determination of the Chief Dimensions.

In Figs. 28 and 29 examples of Koob's method of the graphical determination of the nozzle dimensions and the velocity triangles are reproduced.

Fig. 28 is for a De Laval turbine, Fig. 29 for a three-stage impulse turbine, and the relative pressure temperature and pressure-volume for the three pressure stages are plotted.

The diagrams read with the explanations previously given should contain all that is necessary to make

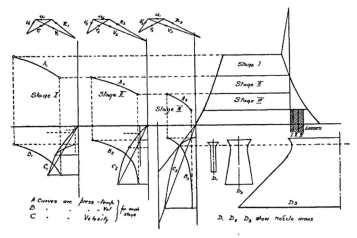

Fig. 29.—Graphical Determination of the Dimensions of Nozzles and the Steam Velocity for an Impulse Turbine with Three Pressure Stages.

them clearly understood (for further details see "Zeitschrift des Vereines Deutscher Ingenieure," Vol. LVIII, Nos. 19 and 21).

Fig. 30 illustrates Bánki's method for the graphical determination of the number of stages in a reaction turbine.

On the line $ab$ as base the curve of inlet velocity is first plotted, wherein the initial velocity $V_1$ of the first stage is taken at 115 to 130 feet per second, and that of the last stage

$$V_z = \text{about } 2u_z \text{ to } 3u_z$$

where $u_z$ represents the peripheral velocity of this stage, $u_z$ can be taken at 330 to 400 feet per second. The energy drop $E_o$ for each pressure stage can now be

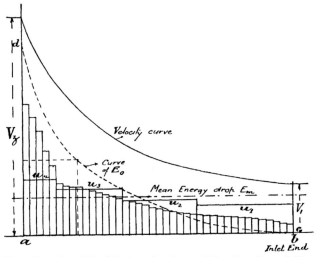

Fig. 30.—Graphical Determination of the Number of Stages in a Parsons Reaction Turbine.

determined from equation (39) if a value is taken for $\delta$ (20° to 25°), bearing in mind that the ratio $\dfrac{u}{V} =$ 0·3 to 0·5 always holds good.

For each point on the base line the corresponding drop is plotted, so that with a large number of stages the curve $E_o$ is obtained, the area $abcd$ corresponding to the total available fall. The mean energy drop

$E_m$ is obtained by dividing the area *abcd* by *ab*, and the number of stages $z$, by dividing the total fall by $E_m$, *ab* being then divided into $z$ parts.

For calculating the length of the vanes, or the sectional area of the passage between them, the specific volumes of the steam in the separate stages must be determined.

# CHAPTER III.

## DETAILS OF STEAM TURBINES.

### 1. Casings.

THE casing in which the rotor and guides are housed can be of simple shape, except in the Curtis turbine, and is made of cast iron or steel.

The surfaces of the joints are fitted by grinding without any special packing, and the casing is

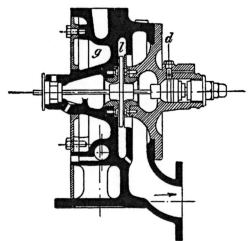

FIG. 31.—Casing of the De Laval Turbine.

usually sheathed with polished sheet metal in order to prevent loss of heat by radiation.

(a) In the *De Laval* turbine (Fig. 31) the casing *g* is closed by a cover *d*, which on being removed

(72)

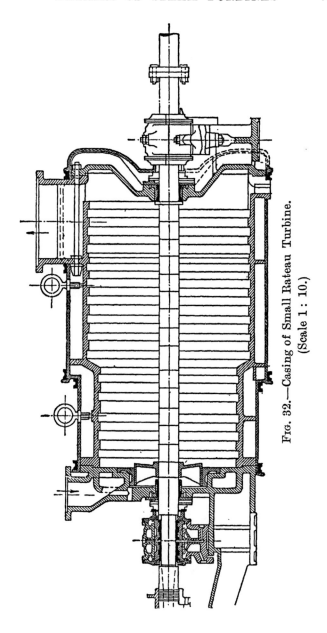

Fig. 32.—Casing of Small Rateau Turbine.

(Scale 1 : 10.)

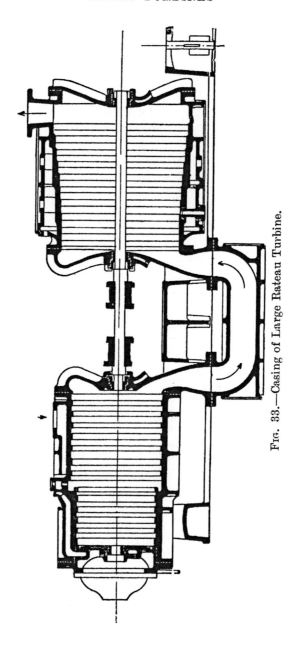

Fig. 33.—Casing of Large Rateau Turbine.

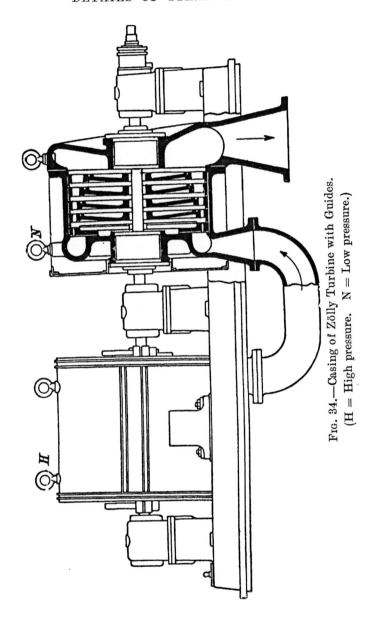

Fig. 34.—Casing of Zölly Turbine with Guides.
(H = High pressure.   N = Low pressure.)

exposes the rotor $l$. The front shaft bearing is arranged separately from the casing, whilst the rear bearing is situated in a conical casting bolted on to the casing proper.

(b) In the smaller sizes of *Rateau* turbine (Fig. 32), all the rotors and guides are enclosed in one casing; but in the larger sizes (Fig. 33) the high and low pressure members are housed in two separate casings, a bearing being arranged between the two in order to shorten the length of unsupported shaft, which therefore rests on three bearings. The casing is adapted to the progressive increase in the diameter of the rotary and stationary members, and its inner surface is provided with turned annular grooves for the reception of the guides. The casing is divided horizontally.

(c) A very similar form of casing is used for the *Zölly* and *Hamilton-Holzwarth* turbines. In the former (Figs. 34 and 35) the supports of the casing are attached immediately below the plane of the joint, so that no appreciable displacement, due to heat, is set up in the horizontal central plane. The thrust on the guides is taken up by a projecting edge in the casing.

In the new patterns the whole of the rotors and guides are housed in a single casing, which is divided horizontally (Fig. 35).

A single casing is also used for small sizes of the Holzwarth turbine (Fig. 36), but for outputs exceeding 750 kilowatts the high and low pressure members are housed in separate casings. A powerful box-shaped foundation, formed of three sections bolted together, supports the casing, bearings, and dynamo.   There is no rigid connection between the casing and foundation at the admission, but at the exhaust end the two are secured together by bearing supports and bolts.

(d) The cast-iron casing of the *A.E.G.* turbines (Fig. 37) is divided, at right angles to the shaft, into

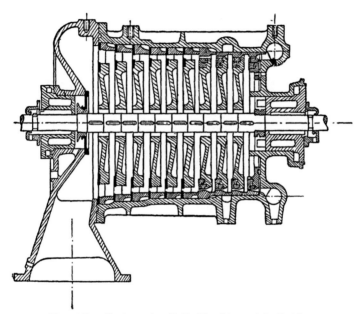

Fig. 35.—Casing of a Zölly Turbine with Guides.

a number of sections corresponding to the pressure stages, and is bolted loosely to the frame, in order to allow free expansion on all sides when the temperature rises.   This also prevents more than a minimum of heat being transmitted to the casing, the contact surfaces between casing and frame being small.   The steam distribution chamber is cast in the casing, access being gained to the interior through a detachable cover at the side. When velocity

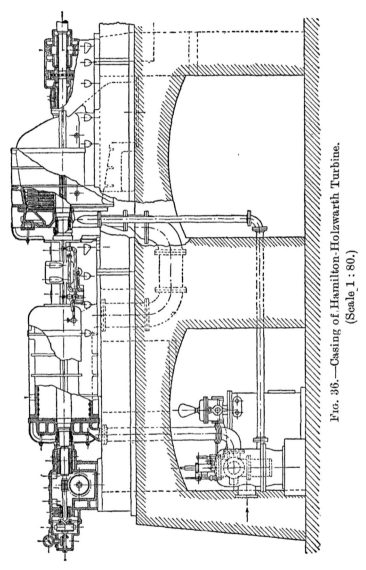

Fig. 36.—Casing of Hamilton-Holzwarth Turbine. (Scale 1 : 80.)

staging is arranged for, the distribution chamber may also be cast in the cover.

To prevent loss of heat by radiation, the casing is lagged with insulating material and aluminium sheathing, or covered with polished sheet iron.

When turbines with more than two rotors are used, the dynamo to be operated is mounted between the high and low-pressure turbines.

(e) In the *Parsons* turbine (Figs. 38 and 39) the

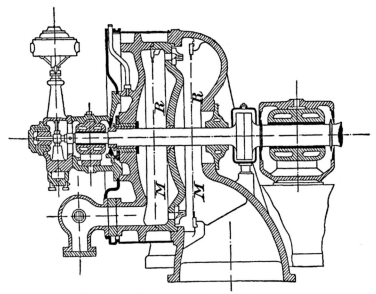

Fig. 37.—Casing of an A.E.G. Turbine.
(M R = Centre line of wheel.)

casing encloses the whole of the rotors and guides, together with the thrust pistons $k$, and is divided horizontally, so that when the upper portion is removed all the moving parts are exposed. For very large powers, such as are used for marine propulsion, the vanes are distributed among as many casings as there are propellors (two to four), the casings then serving to enclose the high, interme-

diate, and low-pressure turbines.   In such cases
the astern turbines necessary for reversing the

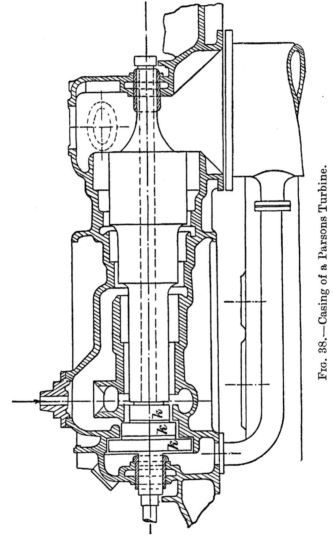

FIG. 38.—Casing of a Parsons Turbine.

vessels are mostly mounted in the casing of the
low-pressure turbines, and on the same shafts as the
latter.   (See Fig. 104.)

At the exhaust end the casing is fastened to the base-plate with heavy bolts.  At the high-pressure end it frequently rests on planed guide shoes, so that it may be able to move in accordance with the temperature of the steam.

(*f*) The upper half of the casing of the *Allis Chalmers* turbine (which is also of the full-admission, reaction type) is mounted somewhat above the centre

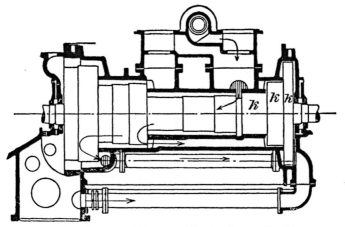

Fig. 39.—Casing of Parsons-Westinghouse Turbine.

of the turbine, in order to prevent the vanes from brushing against the casing even in the event of the shaft bending.

(*g*) The *Sulzer* turbine.— The casing of this turbine is illustrated in Fig. 40 in the high-pressure stage; the turbine works as a partial-admission turbine with velocity stages ; the low-pressure turbine as a full-admission, reaction turbine.

(*h*) The *Curtis* turbine is provided with a very complicated casing (Fig. 41).  The upper part serves to house the dynamo, the central part encloses the rotors and guides, and the base-plate forms a surface

6

condenser so that a good vacuum is obtained and

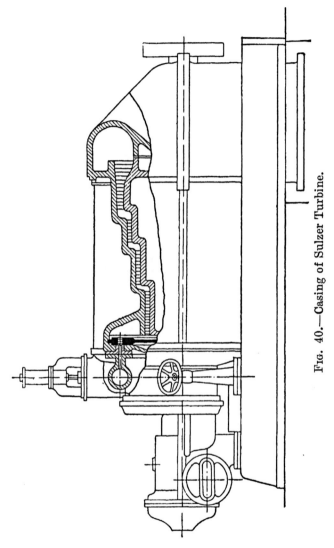

Fig. 40.—Casing of Sulzer Turbine.

space is economised. The governor housing is bolted on to the top of the casing.

(i) The *Melms & Pfenniger* turbine (Fig. 42),

which is a combination of an impulse turbine with a

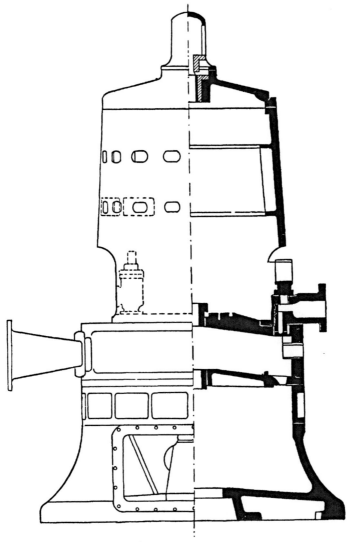

Fig. 41.—Casing of Curtis Turbine.
(Scale 1 : 35.)

full-admission reaction turbine, has a casing analogous

in construction to that of the Parsons turbine, the

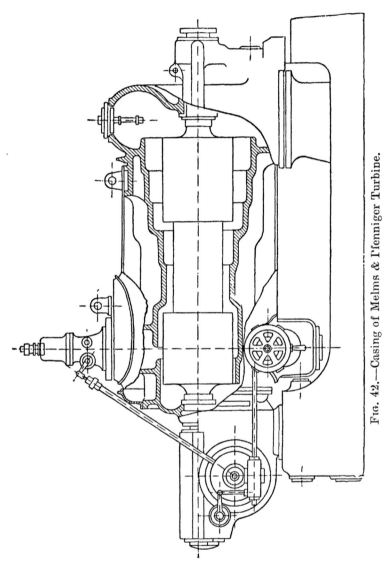

Fig. 42.—Casing of Melms & Pfenniger Turbine.

steam traversing the high-pressure impulse part on its way to the low-pressure reaction part. The smallest diameter is about 1¾ to 2 times that of a purely reaction turbine of equal capacity, the drum of which, moreover, is twice as long. The steam feed pipes are led round the cylinder in order to heat the latter uniformly.

## 2. Nozzles and Guides.

(a) **Nozzles.**—Bronze or steel is the preferred material for the nozzles, these metals being able to resist corrosion and the high temperature of the steam, and taking a good polish. The nozzles are made by pressing the metal into a die, or drawing it over a mandrel. Steel is used more particularly for highly superheated steam, and bronze for wet steam. The flared portion of the nozzle may be of

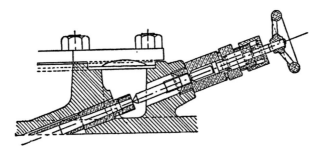

Fig. 43.—Nozzle of De Laval Turbine (Scale 1 : 7).

either circular or rectangular cross section, the latter enabling the steam to impinge on the wheel as a square jet, whilst with the circular nozzles the jet is elliptical. Square nozzles, however, can be set more closely together, so that the action on the rotor is more continuous; though, on the other hand, there is more liability to obstruction of the flow of steam, by contraction of the jet, in the case of square nozzles.

De Laval provides for each nozzle (Fig. 43) a

special throttling spindle which is packed by means of a gland. The nozzles themselves fit in conical bores in the casing and are self-tightening. In order to ensure economical steam consumption under small loads and when running light, De Laval now provides an inner nozzle *d*, which is adjustable in the

FIG. 44.—Modified Nozzle of De Laval Turbine.

main nozzle D (Fig. 44). This arrangement leaves between the two nozzles an annular space, the sectional area of which can be modified by adjusting the inner nozzle. When the turbine is running light, this annular passage can be closed by means of suitable

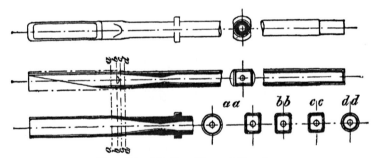

FIG. 45.—Riedler-Stumpf Nozzles.

devices, the steam being admitted through the inner nozzle *d* only.

Fig. 45 represents the square nozzles proposed by Reidler-Stumpf.

(*b*) **Guides.**---Partitions with guide vanes are used in multiple-stage turbines, the rotors of which—each having one or more velocity stages—revolve in

separate chambers in which the pressure is constant.
To this class belong the Rateau, Zölly, and Curtis
turbines.

In nearly all instances the guide vanes are embedded in circular grooves in the casing.

1. *Rateau* employs cast-steel discs *s* (Figs. 46, 47),

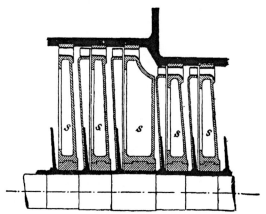

Fig. 46.—Rotors and Guide Discs of Rateau Turbine.

riveted on to cast-steel hubs, the discs being divided
horizontally and made steam-tight by tongue and
groove joints, so that when the
casing is removed the interior is
rendered accessible. The shaft
passes—as far as possible steam-
tight—through the bore of the
rotor hubs. The rotor hub,
which passes through the hub
of the partition, is packed
against the latter by means of
divided bushes lined with white

Fig. 47.—Guide Vanes
of Rateau Turbine.

metal engaging in annular grooves in the hub. The
peripheral portions of the partitions, on which the guide

vanes are mounted, are offset with relation to each other, and the angle of offset is calculated in accordance with the peripheral velocity of the rotors, so that the steam, issuing from the one rotor, enters the succeeding set of guides, instead of impinging on the solid portion of the partition and producing shocks and losses of energy.　In the first stages the admission

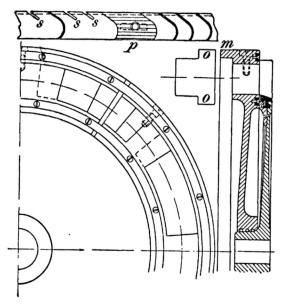

Fig. 48.—Guide Disc of Zölly Turbine.

is only over a portion of the guide vanes, but increases progressively toward the exhaust side, and is finally complete, the diameter of the partitions also increasing at the same time in accordance with the expansion of the steam.

The guide vanes are stamped out of sheet bronze or nickel steel, bent to shape and set in recesses in the partitions, each vane being secured by two rivets.

On their concave side the partitions are covered

with bolted-on discs of sheet iron, and the bores for
the passage of the rotor hubs are made steam-tight,
by divided labyrinths.

In order to facilitate access to the inner parts after
the removal of the casing, the partitions are made in
two halves, joined longitudinally.

2. In the *Zölly* turbine, the projecting rim *m* of

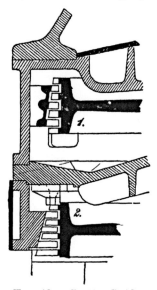

Fig. 49.—Curtis Guides.
(1 = 1st pressure stage.   2 = 2nd pressure stage.)

each guide disc (Figs. 34, 35 and 48) fits close against
the preceding one, the pressure on the discs being
taken up by a projecting edge of the casing. The
upper halves of the cast-steel or cast-iron partitions
are bolted on to the upper half of the casing, and can
therefore be removed along with the latter. The
joint surfaces of the guide discs are ground to fit one
another. The guide vanes are embedded or cast in
the partitions in groups; and between each pair of

groups is a stay $p$, supported by a tongue and groove against an adjacent wrought-iron crown.

The bores in the partitions surround the hubs of the rotors, a steam-tight joint being obtained by means of a labyrinth ; or else bushes, filled with white metal, are provided where the shaft passes through, these bushes surrounding the rotor hubs with a small amount of play, in order to make the joints between the adjacent pressure stages as steam-tight as possible.

The guide vanes are held in position in oblique slits $ss$ in the disc and adjoining ring, by means of projections $o$, and are secured in position at the points where the steam emerges, by rings laid in grooves.

3. In the *Curtis* turbine cast-iron partitions (Figs. 41, 49) are used for the separate pressure stages, which are provided with hand-operated (and also automatically adjustable) valves, to enable a certain difference of pressure to be maintained between the stages—an arrangement that has been found advantageous.

The bronze or cast-iron guide vanes, cast in segments, can be inserted from the outside parallel to the axis of the vertical wheel, so that the free space between the fixed and movable vanes can be modified, the adjustment being effected from the outside by means of lever-operated eccentrics with locking nuts (Fig. 50).

4. The *Holzwarth* turbine, with full admission, shown at the St. Louis Exhibition of 1904 by the Hooven, Owens, Rentschler Co., Hamilton, Ohio, has cast-iron guide discs made in one piece, and the

rotor hubs turn in the bores of these discs (Fig. 51).

The forged steel-plate guide vanes, which are stamped in a cast-iron die, are provided with lugs and are inserted singly in grooves in the guide casing, where they are riveted, the lugs filling up the grooves. When the vanes have been inserted the edges are accurately ground, and a powerful steel ring is shrunk on the outer edge, this ring

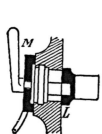

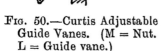

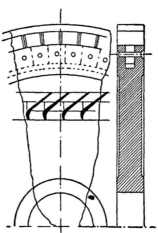

Fig. 50.—Curtis Adjustable Guide Vanes. (M = Nut. L = Guide vane.)

Fig. 51.—Guide Disc of the Hamilton-Holzwarth Turbine.

fitting into a turned groove in the casing of the turbine. The minimum length of the vanes on the admission side is $\frac{1}{8}$th inch.

The construction and form of the guide discs and guide vanes of the reaction turbines correspond exactly to the rotors and rotor vanes of same.

### 3. Rotors and Rotor Vanes.

In all impulse turbines the dimensions of the steam jet must be kept within narrow limits, owing to the great losses due to resistance of flow which occur with high steam velocities; that is to say, the number of vanes per wheel must be increased as the number of stages diminishes. The width of the vanes does not exceed 1 inch except in the low-pressure wheels.

When shrouded vane rims are used, the amount of play between the rotor vanes and the casing may be practically of any desired dimensions.

With open vane rims, the pitch must be smaller in order to lessen the amount of steam spreading in the vanes and endeavouring to escape outwardly. It is often found that the radial height of the vanes is greater than is necessitated by the height of the admission channel.

## (a) De Laval Steam Turbine.

The turbine wheel consists of forged nickel steel and, in the smaller sizes, is pressed on to a sleeve (Fig. 52) which is shrunk on to the shaft. In the larger sizes the wheel is bolted to the flanged shaft ends (Fig. 53). The thickness of the wheel increases considerably towards the shaft, and tapers off and is recessed towards the periphery, so that the stress in the material may be much greater at the periphery than in the other portions of the wheel, and fracture will first occur at the recess. Nevertheless the factor of safety is about 2 at the periphery of the wheel. According to calculations, fracture will not occur until double the normal speed is attained. Fracture of course destroys the equilibrium of the wheel, and the hub is so close to the casing that, in these cir-

cumstances, it comes in contact therewith and a
braking effect is produced (Fig. 31).

The profile of the wheel is a logarithmic curve,
which is asymptotic with relation to the radial axis
of symmetry.

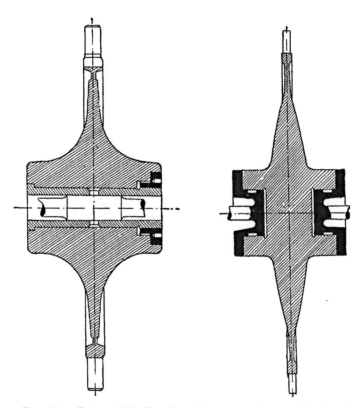

FIG. 52.—Rotor of De Laval        FIG. 53.—Rotor of De Laval
        Turbine.                              Turbine.

The vanes (Fig. 54) are forged in dies. They are
milled on both sides, inserted interchangeably in the
wheel rim and caulked. Lugs provided at the periphery of the vanes form a closed ring all round. .

## (b) Rateau Turbine.

In the Rateau multiple-stage impulse turbines the rotors (Fig. 55) are made of open-hearth steel, and are run at a peripheral velocity of 330-460 feet per second. They are pressed into a slightly conical shape. In

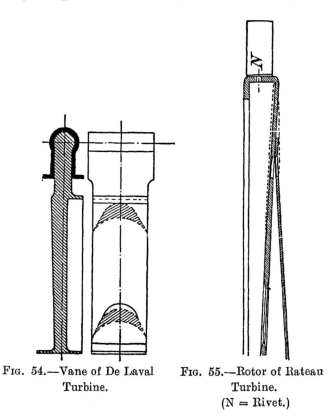

FIG. 54.—Vane of De Laval Turbine.

FIG. 55.—Rotor of Rateau Turbine.
(N = Rivet.)

the high-pressure rotors the outer edge is bent round cylindrically, whilst the edges of the low-pressure rotors are bent over twice at right angles. The high-grade nickel steel vanes are made in pairs by passing the suitably cut strips of sheet metal through a drawing press. They are mounted astride the periphery of the disc, and are riveted at the side, so that no obstructive rivet heads come in the steam passages.

A steel band is riveted to the outer edge, to impart rigidity and prevent waste of steam by leakage.

In a wheel 45 inches in diameter, for example, built to run at a peripheral velocity of 310 feet per second, the number of vanes will be 240 and their width $\frac{3}{16}$ of an inch.

The distance between the rotor and the guide vane crown is about $\frac{5}{32}$ to $\frac{7}{32}$ of an inch, and a similar interval is provided between the rotor and the casing.

The wheels are light in weight, and remain in equilibrium at far higher than the contemplated peripheral velocities.

### (c) Zölly Turbine.

In the Zölly turbine, in order to prevent rust and steam friction, the vanes are milled from highly polished nickel steel or made of drawn and stamped bronze. They are tapered outwardly, so that their radial length can be increased, whilst the stress on the material is uniform throughout, and the low tensile stress at the root of the vanes gives a high factor of safety.

In consequence of the uniform strain over the whole length of the vanes, the peripheral velocity can be increased and the number of stages diminished, so that the length of the turbine and cost of construction are lessened. In order to allow for the expansion of the steam jet between the vanes and enable the steam to be properly guided, the surface of the spacing pieces between the vanes is curved (Fig. 56). The periphery of the rotor vanes is not surrounded by a steel band in all cases.

The rotor discs (Fig. 38) keyed on the shaft are forged in one piece out of polished open-hearth steel, and accurately balanced. A riveted, polished steel covering forms a groove (Fig. 56) at the rim of the disc, for the reception of the vanes and spacing pieces.

FIG. 56. — Rotor Vane of the Zölly Turbine.

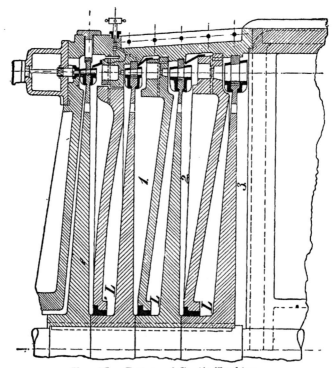

FIG. 57.—Rotors of Curtis Turbine.

L = Rotor.   1 = 1st pressure stage.   2 = 2nd pressure stage.
3 = 3rd pressure stage.

### (d) Curtis Turbine.

The arrangement of the rotors and the method of securing the segmental vanes of same are illustrated in Figs. 57 and 58. The vanes are made of hard bronze, and a bronze ring is riveted on to each row. The axial play varies between $\frac{1}{10}$ and $\frac{7}{32}$ of an inch, the larger value being for the low-pressure stages. The rotors are usually made in one piece and are thickened toward the hub. Latterly,

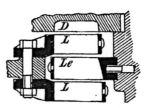

FIG. 58.—Method of Securing the Vanes in the Curtis Turbine. (D = Nozzle. L = Rotor vane. Le = Guide vane.)

however, the rotors have been made of two plates of sheet metal, riveted to a common hub, and carrying peripheral vanes secured by riveting. In small rotors the vanes are milled out of the solid rim. In some cases the vanes are also mounted in a special manner (Fig. 58).

### (e) Holzwarth Turbine.

In the Holzwarth turbine steam is admitted all round the rotors, the construction of which is shown in Fig. 59. The body of the rotor is formed of two discs of steel plate, riveted to a cast-iron hub. The discs taper outwardly, are bent outward at the edges, and are keyed on to the shaft. The vanes, of pressed sheet steel, are riveted, and surrounded at the edges by a steel band. In order to reduce their weight the vanes are made hollow, and they are turned so as to fit the cylindrical casing of the rotor exactly.

### (f) A.E.G. Turbine.

In the A.E.G. turbine (Fig. 37) the rotors (Fig. 60) are of nickel steel milled to a section of uniform strength from the solid metal. They are separated by an intermediate cover and mounted in a common casing. The vanes, drawn from tough bronze, are secured by dovetail joints in grooves in the wheel

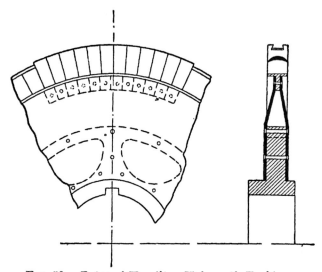

FIG. 59.—Rotor of Hamilton-Holzwarth Turbine.

rim. On the outer side they are enclosed by light strips of steel or bronze.

The rotors are secured to the shaft by tapered bushes.

### (g) Parsons and Parsons-Westinghouse Systems.

The guide vanes (Fig. 61, a) are secured directly to the inside of the casing, and the rotor vanes are mounted—in the spaces between the guide vanes—

on the periphery of a steel drum slipped over the shaft, the drum in large sizes being connected with the shaft by cast-steel wheels.  A drum of this kind, $8\frac{1}{2}$ feet long and $10\frac{1}{2}$ feet in diameter, is made of $2\frac{1}{2}$ inch metal.

In smaller sizes the hollow drums are forged, the

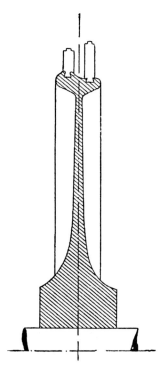

Fig. 60.—Rotor of A.E.G. Turbine.

metal of one 10 feet long and 55 inches diameter being 2 inches thick.  For large units the drums are generally built up of sections.

The rotor and guide vanes, which are of very light construction, are rolled or cold-drawn into long strips, and cut into suitable vane lengths.  In order to enable the vanes of the first stages to withstand the high steam temperature for a longer period, they are made of an alloy specially rich (up to 98 per cent) in copper, those of the other stages being made of a special malleable bronze. Alloys containing 16 parts of copper and 3 of tin, or 80 per cent copper and 20 per cent nickel, are frequently used; and in some new patterns the vanes are of drawn steel.

The material used has a tensile strength of 74,000 to 79,600 lb. per sq. inch, with an elongation of 20 to 30 per cent, and the dimensions ensure a high factor of safety. The pressure of the individual vanes in such case is very small, and the amount

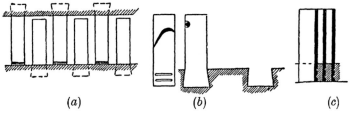

     (a)          (b)          (c)

FIG. 61.—Guide and Rotor Vanes of the Parsons Turbine.

of play between the guide and rotor vanes is never less than $\frac{1}{8}$ inch, increasing to 1 inch for the long vanes of the low-pressure side. The vanes are secured in the casing and rotor by means of turned slightly dove-tailed grooves (Fig. 61, b), bronze spacing pieces being inserted in the grooves, between the vanes, and caulked. The foot of each vane is provided with two parallel semi-circular grooves into which the material of the spacing pieces is forced by caulking (Fig. 61, c). Vanes projecting more than 4 inches are stiffened by means of narrow rings—of bronze wire soldered with German silver—partially inlet into the edges of the vanes a little below the free end, the bronze wire being secured to each vane by lapping it with thin copper wire.    Since with this

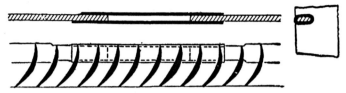

FIG. 62.—Stiffening the Vanes of the Parsons Turbine.

method of attachment the high soldering temperature is liable to injure the structure of the material of the vanes, Parsons employs a system of binding, instead of soldering, the wire having a cross section in the

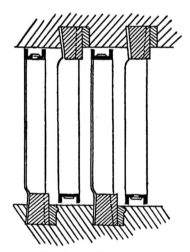

FIG. 63.—Method of Attaching the Vanes in the Allis-Chalmers Turbine.

shape of a comma, and being fitted into a corresponding recess in the vane.

The low-pressure vanes of large turbines are stiffened by means of brass strips (Fig. 62) of oval section, soldered in recesses in the vanes and lapped with copper wire. To allow for expansion, these strips are cut through at different points on the periphery, the ends being spaced and enclosed in short metal tubes.

The maximum thickness of the vanes is $\frac{1}{21}$ to $\frac{1}{5}$ of an inch, and the width parallel to the axis $\frac{1}{4}$ to $\frac{1}{2}$ inch, the longer vanes being correspondingly broader. In turbines of about 1000 h.-p., the spacing of the vanes in the first stage is $\frac{1}{5}$ of an inch, the axial height of the vanes $\frac{1}{3}$ to $\frac{1}{2}$ inch, the radial length $\frac{1}{2}$ to $\frac{3}{4}$ inch ; in the final stage, spacing $\frac{3}{4}$ inch, axial height $\frac{1}{8}$ inch, radial length 6 inches.

In the Parsons turbines, built by Messrs. Willans & Robinson, two-part rings, in which the vanes are secured, are inserted in the drum and casing.

As the diameter of the drum diminishes, the height of the vanes increases about in proportion to the square of the difference in diameter ; and in consequence of the resulting lessened peripheral velocity, and the necessarily diminished velocity of the steam, the number of rows of vanes must be increased, the drum being therefore made longer.

The entering angle of the vanes is about 65 to 70°, and the issuing angle is about 20°.

### (h) Allis-Chalmers Turbine.

In the Allis-Chalmers turbine the roots of the vanes are dovetailed, and are inserted in milled grooves provided in separate bottom rings (Fig. 63). These rings are also dovetailed, fit in grooves in the drum and casing, and are held in position by caulked locking pieces. The vanes are stiffened by means of a peripheral band of channel section, which has very thin flanges and is riveted to the vanes by lugs. This precaution greatly reduces the risk of vibration in long vanes.

### (i) Strength of Rotor Discs and Drums.

The influence of the bore of the boss on the stress to which high-speed discs are subjected is extremely great. The bore doubles the stress in comparison with solid discs, the amount of the stress decreasing from the centre towards the periphery. Consequently, the centrifugal force of the disc and of the vanes secured thereon must be considered separately.

If $u$ represents the peripheral velocity, then the stress in an iron disc due to centrifugal force may be calculated approximately as follows :—

$$f = \text{approx. } 0.105 \ u^2 \text{ lb. per sq. inch.} \qquad (49)$$

Presuming the vanes to be of constant section, the stress on the root of the vanes will be

$$f_1 = \frac{4\pi^2 \cdot \rho D l n^2}{g} \text{ lb. per sq. inch} \qquad (50)$$

wherein $\rho$ denotes the weight per unit length, $l$ the radial length of the vanes, D the diameter of the wheel, and $n$ the number of revolutions per minute.

The vanes and the attachments of same are assumed to be uniformly distributed over the periphery of the wheel.

If $y_n$ represent the thickness of the disc at the centre of the shaft, $u$ the peripheral velocity at the distance $x$ from the axis $= \omega x$, $\rho$ the density and $e$ the basis of the natural or hyperbolic logarithm, then the accurate formula for estimating the thickness of a high-speed disc, of uniform strength, at the distance $x$ from the axis, will be

$$y = y_n e^{-\left(\frac{\rho\omega^2 x^2}{2fg}\right)}$$

$$= y_n e^{-\left(\frac{\rho\omega^2}{2fg}\right)} \qquad (51)$$

$f$ = stress in material

The calculation for high-speed drums can be based on the assumption that they consist of a number of adjacent revolving rings.

### SHAFTS.

(a) The shafts carrying the rotors are made of open-hearth or nickel steel, and are frequently reduced in diameter at the ends (Fig. 36), in order to facilitate threading the wheels.

The drums of the Parsons turbine are generally of cast steel, or else the ends alone are cast-steel cylinders, the central shell consisting of a drawn tube of mild open-hearth steel. Seamless rolled steel shells are also used.

The shafts must be regarded as resting freely on the bearings. With $l$ representing the length in inches between bearings, and $n$ the number of revolutions per minute, the diameter D in inches of the shaft may be calculated from the formula

$$D = \frac{nl^2}{48 \cdot 2 \times 10^6} \qquad (52)$$

A flexible coupling is interposed between the turbine and the dynamo.

## Critical Speeds.

(b) For the speeds at which turbines are now run —except in the case of the De Laval turbine—flexible shafts, which are self-adjusting in the axis of gravity, are no longer used. The only kind suitable is the rigid shaft, run at less than the critical speed, modern workmanship giving a sufficiently accurate balance of quick-running shafts and discs. Hence in calculating the diameter of shafts it is sufficient to take bending stresses and deflection into consideration.

In the case of shafts running at very high speeds, however, the critical speed must be ascertained, and the shaft must be run at a number of revolutions sufficiently below that limit, because the whirling of the shaft may increase, at the critical speed, to a degree resulting in fracture, and the running will be very uneven.

When an unbalanced disc and shaft revolve, their common centre of gravity does not coincide with the axis of the shaft, but lies (e.g.) in the axis CD (Fig. 64). The excess of centrifugal force in the rotating mass causes the axis to become displaced on the more heavily loaded side of the disc, so that the geometrical centre of the disc describes a circle a (Fig. 64). To ascertain at which point of the disc material must be added or removed in order to restore the equilibrium, all that is needed is to hold a piece of chalk, for instance, in such a manner that it just touches the periphery of the revolving disc at the point of maximum displacement, a trial weight being then attached to the disc at the point diamet-

rically opposite to the chalk mark, to see if it bal-
ances the disc.  When the critical speed is attained,
a sudden powerful vibration is produced, quiet running being restored when this speed is exceeded.
This vibration is caused by the shaft and disc leaving their geometrical axis of rotation and revolving
about the axis CD, the prolongation of which passes
through the bearings, whilst the shaft bends to such
an extent that its geometrical centre describes the

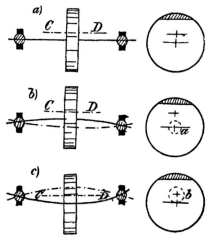

FIG. 64.—The Whirling of Flexible Shafts.

circle *b* about the centre of gravity, so that a chalk
mark indicating the position of the requisite balance
weight would be on the side of lowest load of the
disc.

Now, since the shaft and disc have a natural
period of oscillation, just like a spring or a pendulum,
the change in the axis of rotation occurs when the
time of rotation coincides with the period of natural
oscillation.

In the De Laval turbine the critical speed is about

$\frac{1}{5}$ to $\frac{1}{8}$ the normal speed, so that this critical speed is exceeded during the starting phase. This must be accomplished quickly. If such a shaft be mounted so as to be self-adjusting and capable of giving, the shaft rotates about a free axis of rotation, which passes through the centre of gravity, when running at a multiple of the critical speed.

The dimensions of the shaft are frequently such that the deflection produced by the stationary load is greater than the critical deflection.

If I be taken to represent the Moment of Inertia, W the weight of the shaft and disc in lb., $l$ the length between bearings in inches, and $n$ the critical number of revolutions per minute, then

$$n = a \sqrt{\frac{I}{Wl^3}} \qquad (53)$$

in which for shafts freely supported on both sides

$a =$ may be taken as about 420,000

and for shafts held tightly at both ends

$a = 840,000$ (about).

If the critical velocity be above the working velocity, that is to say, only a slight set of the shaft is permissible (about $\frac{1}{500}$ inch), the shafts will be too small. In the case of extensive set, the critical velocity is frequently below the working velocity.

### Balancing the Rotating Parts.

The rotating parts must be balanced in one of two ways :—

(a) In the case of wheels, static balancing—which can be effected with considerable accuracy—is usually sufficient.

(*b*) Dynamic balancing, in the case of drums, is usually effected by the aid of rotary devices provided with spring bearings and displaying the lack of balance in exaggerated form, by means of pointers.

The best method is to balance each drum separately at first, then again after it has been fitted with the vanes, and finally when it is in position in the casing.

Special difficulties are presented by drums coupled to heavy dynamo armatures, since the windings of the armature are very liable to become shifted in position under the influence of centrifugal force.

Drums are liable to undergo changes of shape in consequence of the increase of temperature, particularly in the early stages of running. Hence, in many cases, the final turning down is postponed until the drum has been run for about twenty-four hours under full steam pressure. Increases up to about $\frac{1}{6}$ of an inch in the diameter of the drum after prolonged running are not infrequent.

### Bearings.

(*a*) All turbine bearings have to stand high journal velocities (16 to 50 feet per second), in consequence of which the friction is heavy, increasing as the square root of the third power of the peripheral velocity of the journals.

The amount of heat-energy E, given out by bearings of given length *l*, and diameter *d*, will be

$$E = atdl \text{ per hour} \qquad (54)$$

in which *t* indicates the difference between the temperature of the bearing and that of the outer air,

$d$ and $l$ are the length and diameter of the journal and $a$ is a constant.

In general the following formula will be applicable to turbine bearings, if $t$ be taken to represent the temperature of the surfaces in sliding contact and W the load on the bearings,

$$W t = \beta dl \qquad (55)^{1}$$

where $\beta$ is another constant.

(b) The type of bearing used must meet the requirements of high speed on the part of the shafts and journals, the behaviour of the shafts being an important factor.

For horizontal shafts which may be regarded as perfectly rigid, and for which the critical speed is a negligible factor, ring-lubricator bearings with two to four lubricating rings, and with spherical working faces to the cups—similar to Sellers' bearings—are chiefly used. Water cooling is largely employed.

The cups of Sellers' bearings are made in two parts, held together by screws. For small sizes the cups, which are turned down cylindrically on the outer side, may also be of bronze. It is advisable to arrange the bearings quite separately from the casing, in order to minimize the transmission of heat from the casing to the bearing; this is done in the Zölly turbine, which is fitted with a simple bearing —faced with white metal and lubricated under pressure—on the high-pressure side, and with ball bearings on the low-pressure side.

---

[1] A rather more useful rule is that the bearing pressure in lb. per sq. inch multiplied by $u$ the peripheral speed of the bearing in feet per second = about 2500.—[Ed.]

The main bearings of the Parsons turbine are all filled with white metal.

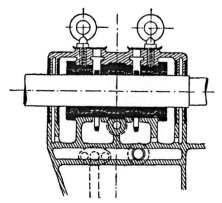

Fig. 65.—Bearing with Ring Lubrication.

Similar bearings for Rateau and Parsons turbines, for driving dynamos, are illustrated in Figs. 65 and

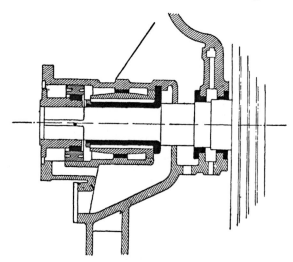

Fig. 66.—Bearing of a Parsons Turbine.

66. The bearings are cast on the casing. The tubes shown in Fig. 65 are for cooling water.

In machines with an output exceeding 1000 kilo-watts, the sliding speed of the journals is so small that elastic bearings are unnecessary.

For smaller machines Parsons uses a bearing which enables the rigid shaft to be adjusted in the neutral axis by the " give " of the cups (Fig. 67). These consist of a number of thin-walled, undivided bronze bushes, slipped one over another, with a play of about $\frac{1}{25}$ of an inch, and protected from twisting by means of cottar pins. A layer of oil between the sleeves enables the shaft and rotors to be adjusted in

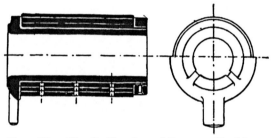

Fig. 67.—Elastic Bearing of Parsons Turbine.

the neutral axis. The innermost bush is bored out slightly eccentric, to facilitate exact adjustment with regard to the casing. The oil between the concentric sleeves circulates by capillary action, and is forced in through bores in the sleeves. For higher outputs, end bearings of the kind shown in Fig. 66 are used.

The difficulties encountered in mounting bearings that run at very high speeds were overcome by De Laval by arranging for a certain amount of " give " on the part of the bearing, although the shaft itself is rigidly supported by the bearing. A terminal ball-bearing of a De Laval turbine is illustrated in Fig.

68. The bearings are lined with white metal and provided with spirally arranged oil-grooves; they are adjustable, and are guided by a special packing gland at the outer end.

The A.E.G. makes its turbine bearing cups of cast iron with white metal, and provides water cooling. The axial thrust occurring in marine propulsion is taken up by the aid of collar thrust bearings, into which oil is pumped under pressure by a valveless rotary pump.

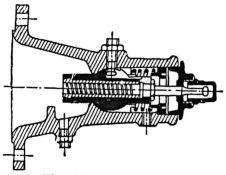

Fig. 68.—Thrust Bearing of De Laval Turbine.

(c) Although in reaction turbines (Parsons and Parsons-Westinghouse) the axial thrust is taken up by special dummy pistons, they are always provided in addition with a collar thrust bearing at the end, analogous to the thrust block of the marine engine. This device (Fig. 69), which is usually made in two pieces, can be adjusted by means of set screws so that the axial play is reduced to about $\frac{1}{100}$ of an inch. The usual arrangement is to make one half of the bearing adjustable inwardly and the other half outwardly, so that the collars make contact on either side. The corresponding adjustable bearings in the

Zölly turbine are chiefly to maintain the distance between the guides and rotor about $\frac{1}{8}$ of an inch apart.  If the thrust bearing be screwed up too tight,

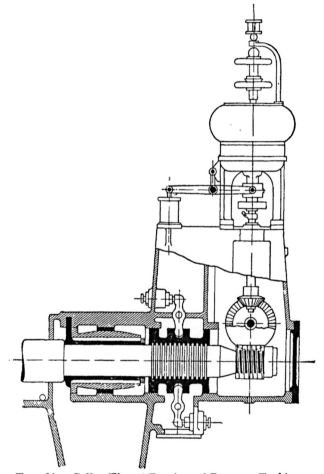

Fig. 69.—Collar Thrust Bearing of Parsons Turbine.

considerable vibration will be set up at low and high speeds.

(d) Turbines with vertical shafts (Curtis system) require footstep bearings.  In recent patterns of this

8

class of turbine, this bearing (Fig. 70) consists of
two cast-iron blocks, one supporting the end of the
shaft, whilst the other can be adjusted vertically by
means of a powerful screw. Oil or water is forced
down into a recess in the lower block, under sufficient
pressure to take up the weight of all the moving parts.
If oil be used the bearings are lined with white

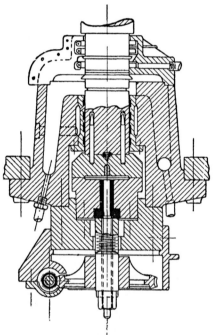

Fig. 70.—Footstep Bearing of Curtis Turbine.

metal, but in the case of water wooden linings are
used. The whole of this bearing rests on a timber
foundation, and the bearing itself is surmounted by
a collar thrust bearing. The admission of air is
prevented by making a tight joint between the oil
chamber and the base containing the two bearings,
by means of packing which is acted upon by steam

pressure. In order to obviate the unfavourable results that would ensue from failure of the lubricating device a powerful brake is provided, which fits against a ring on the inner wheel, so as to take up the weight of the moving parts. The brake serves also to stop the turbine, which would otherwise continue running for some considerable time.

It is desirable to omit any stopcock on the oil-feed pipe, and by careful supervision to prevent waste of oil.

### (e) Lubricating Arrangements.

Owing to the high velocity in the bearings, special care must be given to the lubricating arrangements, and oil pumps are therefore essential.

1. In the *Parsons* turbine the main bearings are lubricated automatically, the oil being forced by a special pump (Fig. 97) through the bearings at a pressure of 17 to 20 lb. per sq. inch, after which it is cooled by passing it through a coil immersed in water, before recommencing its cycle. The bearings are lubricated by means of a plunger pump before starting the turbine.

A mixture of valvoline cylinder oil and Russian mineral oil is said to answer well. The oil flows to the pumps from the bearings by way of a continuously rotated cooling drum; it remains in fit condition for use up to about two months, at the end of which time it is drawn off and filtered, the shortage being replenished by fresh oil. The consumption of oil amounts to about one half that in reciprocating engines of the same power, one half being cylinder oil. The steam working inside the turbine is not lubricated.

2. The *Rateau* turbines built by the Oerlikon Co. are fitted with a device which stops the turbine in the event of an interruption in the supply of oil, in order to prevent the shaft from getting heated.

3. In a similar manner the oil pump in the *Zölly* turbine is driven from the main shaft by worm gearing, the effluent oil running down into a well in the base-plate. From this well the oil is drawn by a two-stage centrifugal pump, and delivered through a copper cooling coil, under a pressure of about 20 lb. per sq. inch, to the various centres of distribution. In the collector itself the oil is cooled by double cooling pipes traversed by water.

4. The footstep bearing of the *Curtis* turbine is lubricated with oil forced in under pressure, the consumption being about 5 gallons per minute for a 5000 kilowatt turbine; the effluent oil passes into the collar-thrust bearing above the footstep bearing. More recently oil has been replaced by water for this purpose, no special packing being then needed.

5. The bearings of the *A.E.G.* turbines are lubricated with oil forced in under pressure by a valveless rotary pump driven by the turbine itself.

6. In the *De Laval* turbine it is impossible to prevent shaft vibration, and this is transmitted to the lubricating rings of the bearing, so that in some cases they can no longer follow the shaft and therefore lose their efficiency to some extent. Since at the same time the oil gets too hot with ring lubrication, the sight-feed system of lubrication is found more suitable. The lubrication of the rotor shaft must not be interrupted, or the bearings will soon run hot. Wick lubrication, in which the oil drips from the

wicks, is found to be the best, the filtration of the oil being then effected in the wick itself. Ordinary sight-feed lubricators frequently get out of order, owing to the small bore (entailed by the small number of drops) being choked by tiny impurities, whereupon the bearing gets hot at once.

7. In the *Hamilton-Holzwarth* turbine a portion of the base-plate is shaped into an oil tank, from which the oil runs to a pump that forces it through the bearings, after which it flows back into the tank again, passing through a filter on the way. A small valve is provided in the oil feed pipe of each bearing so that the oil supply can be controlled.

8. For lubricating the bearings, *Melms and Pfenniger* provide an oil pump mounted on an auxiliary shaft at right angles to and driven from the main shaft by worm gearing. This pump draws the oil from a tank in the base-plate and delivers it to all the bearings.

### Stuffing Boxes.

It is a very difficult matter to pack the shaft at the point where it issues from the casing of the turbine. The reciprocating motion which has a cooling action in the case of the piston engine is here lacking, and it is necessary to prevent the penetration of air, especially on the vacuum side.

(a) In most patterns the method of packing adopted is the labyrinth system, in which the packing does not come in contact with the shaft, the steam being throttled in the intermediate spaces by the resistance set up by friction and motion.

Parsons was the first to introduce an effective

labyrinth packing (Fig. 71).   Its efficiency is based
on the packing of the steam in the labyrinth with
practically no friction, the metal rings revolving past
one another with a play of about $\frac{1}{250}$ of an inch.

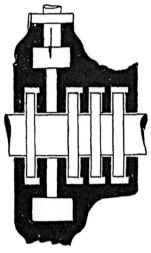

FIG. 71.—Labyrinth
Packing.

A number of concentric
grooves are turned in the
shaft, whilst a divided packing box of special bronze in
the casing is provided with
corresponding projections
fitting with a small amount
of play in the shaft grooves.
The steam entering the labyrinth revolves with the shaft,
and is driven by centrifugal
force against the walls of
the packing box.   In consequence of the numerous
compulsory changes of di-
rection the entering steam loses its pressure, and
also meets with resistance from the high density of
the steam compressed by the centrifugal force.   On
the side nearest the condenser a tight joint is maintained by the aid of steam which is usually drawn
from the exhaust steam of the regulating device, the
admission part being adjusted in such a way that
the steam condenses immediately on issuing from the
labyrinth.

Of late, packing has been effected by means of
small quantities of water, taken from the delivery
pipe of the feed water pump and returned to the
suction pipe.   By the aid of this method of packing a
95 per cent vacuum can be maintained.   The latest

pattern of the labyrinthine projections in the Parsons turbine is comma shaped.

Fig. 72 shows the stuffing boxes used in the ocean greyhounds, " Carmania " and " Lusitania," $k$ representing flexible Ramsbottom rings, $g$ the collecting chamber, $h$ the pipe leading to the auxiliary condenser. The space $o$ connects the high-pressure turbine with the low-pressure turbine, and in this space a pressure of about 3 lb. per square inch is maintained.

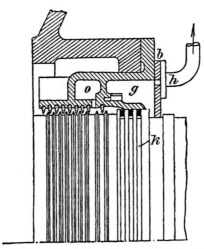

FIG. 72.—Stuffing Box of a Parsons Marine Turbine.

Fig. 73 illustrates the forward stuffing box of a high-pressure turbine ; $a$ being the oil catcher, $b$ and $c$ rings (fifteen to twenty in number) on the shaft, $d$ the stuffing box, $y$ the bronze Ramsbottom rings pressing elastically against $d$, $g$ represents the steam passages. The pressure in $b$ is about 21 lb. per square inch, in $c$ 10 lb. per square inch, in $x$ about 5 lb. per square inch ; $h$ is the labyrinth packing.

Shaft packing by the aid of water has recently

been adopted, this method being used for the dummy pistons shown in Figs. 38, 39. The steam consumption with this form of packing amounts to about 1 per cent of the total.

Similar packing is employed in the Westinghouse turbine, whereas Zölly uses Schwabe's metallic packing.

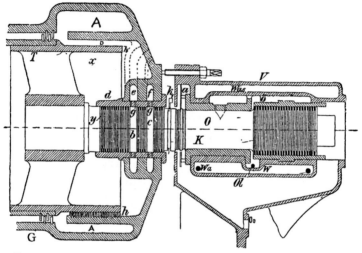

Fig. 73.—Stuffing Box of Parsons Turbine, High-Pressure Side. (A = Atmos. pressure. D = Packing omitted. T = Drum. V = Front. Was = Water jacket. O = Oil passage. K = Collar-thrust bearing. Wa = Water inlet. W = Water outlet. Ol = Oil inlet. Oo = Oil outlet. G = Casing.)

(b) Rateau (Fig. 74) uses tight-fitting aluminium stuffing boxes, which form a labyrinth packing with the two end rotors, and are capped on the outside so as to make an enclosed chamber. Steam under a pressure of 17 to 20 lb. per sq. inch is admitted into these chambers, entering that on the high-pressure side first, and passing thence to the one on the low-pressure side.   The pressure of the steam is controlled by a special regulating device.

The outermost stuffing boxes are formed of a metal packing consisting of a three-part cylinder of grey cast iron, which is pressed against the shaft by coiled springs, and against the stuffing box by means of screw springs and an interposed iron plate.

(c) In the Sulzer turbines, two-part stuffing boxes

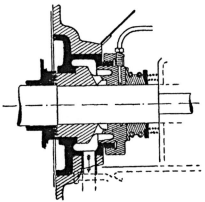

FIG. 74.—Rateau Stuffing Box.

of flexible brass plates are used, the plates being separated by smaller, but somewhat thicker, bronze rings and making a tight joint round the shaft. Water is admitted at the central portion of the low-pressure stuffing box.   Where the shaft issues from the stuffing box the joint is kept tight by the brass plates and a steel bush.

(d) In the Hamilton-Holzwarth turbine the steam is throttled by means of a divided slit of sufficient length, the outwardly forced steam preventing any further steam from entering the slit.

A ring r (Fig. 75) is secured on the shaft, rotating with the latter and engaging in the groove of a second

annular bush *b*, which is secured in the turbine casing
and presses against the end faces of the bearing cups
*l*. Packing is inserted in the recess *o* and pressed
home by the nut *m*. Water is admitted into the resulting cavity from the low-pressure side.

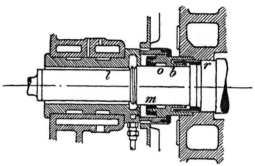

Fig. 75.—Hamilton-Holzwarth Stuffing Box.

(*e*) In the De Laval turbine, divided bushes filled
with white metal and provided with a ball joint and
pressure springs are used, the actual packing medium
consisting of a layer of lubricating oil drawn in toward
the vacuum chamber.

### Governing Steam Turbines.

(*a*) In governing turbines means must be adopted
to enable the engine to work under varying duty.
In the case of impulse turbines this result is obtained
without varying the speed, whilst in reaction turbines
the speed is varied.

In the former case, when, for instance, eight different outputs are to be provided for, the steam required for the highest duty is distributed among eight
nozzles, and the number of nozzles allowed to work
at any one time is varied according to the output desired. In multiple-stage turbines this applies to

each stage.  In these circumstances with a constant fall in temperature the steam velocity and peripheral velocity remain constant.

The majority of governing systems employed work by throttling the steam or by admitting in gusts or puffs.  For reaction turbines these are the only practicable methods, whereas in the case of impulse turbines, governing may also be effected by varying the expansion of the steam, that is to say cutting off a portion of the admission.  It is true that throttling always destroys a portion of the useful energy of the steam, but the method has the advantage of simplicity, and the consumption of steam is not adversely affected under varying loads.  In all systems very effective governing is obtained.

(b) In the De Laval turbine the throttle valve in the steam inlet pipe is controlled by a centrifugal governor.  In this case the throttle valve closes sufficiently to prevent the turbine from racing with reduced load, which may happen, however, in condensing turbines (with high vacuum) when running light, even if the governor throttles the pressure down to that of the atmosphere, since the resistances are very low.  In such case the vacuum in the casing is reduced by means of an auxiliary governor, which admits air into the casing—and so destroys the vacuum—after the throttle valve on the live steam pipe has been closed.

The two semi-cylindrical governor weights b (Fig. 76) are mounted on knife edges a in a casing which is pressed against the end of the intermediate shaft, these weights pressing against a spring plate d and the lever l by means of studs.  The springs can be adjusted by the set screws $m$, and the weights take up the weight of the rotated governor body. As the velocity increases the weights fly outward, the seat of the spring is pressed by the studs, and the governor spindle situated inside the spring is displaced until the admission of steam is correspondingly diminished by the lever $l$ acting on the admission valve.

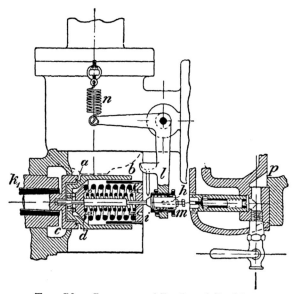

Fig. 76.—Governor of De Laval Turbine.

If the turbine is working with condensation, and the throttling is insufficient, the movement of the governor spindle operates a piston $h$, its position with regard to the bell-crank lever $l$ remaining unaltered, because the spring surrounding the piston $h$ is stronger than the spring which keeps the governor spindle in contact with it, and so keeps the admission valve from closing.

If the load be suddenly taken off, the admission

valve is almost completely closed, because the governor flies open; and, in addition, the valve for admitting air through $p$ into the turbine casing is opened, since the piston is pressed forward by the lever $l$, which is now at rest. If the vacuum has to be maintained for other engines, air enters through a governing device situated in the exhaust pipe of the turbine. The connection between the condenser and the casing is partially interrupted, the pressure in the casing rises and thus increases the resistance of the rotor, so that the expansion of the steam in the nozzles is restricted.

A recently introduced automatic device for shut-

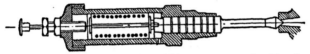

Fig. 77.—Automatic Device for Closing Down the Nozzles of the De Laval Turbine.

ting off the nozzles is shown in Fig. 77. For high outputs the spring is closed by the pressure of steam on the spindle, but as the load decreases the orifice of the nozzle is narrowed by the action of the spring, when the governor begins to throttle the steam.

(c) In the Zölly turbine a relay governor is used, the controlling valve $m$ (Fig. 78) of which is operated by the governor and leads from the pipe $a$ to an oil or water accumulator, in which pressure is generated by a small rotary pump, driven from the turbine by worm gearing, whilst the return pipe $b$ leads to the suction chest of the pump. Pipes $e$ and $f$ connect the valve $m$ with a cylinder mounted on the casing of the balanced governor valve $k$. As the speed increases in consequence of diminished load,

the entrances to the tubes *a* and *f* are exposed, and
communication is set up between *e* and *b*, thus causing the valve *k* to be lowered.   The steam pressure
then declines in consequence of the piston valve *k*
being lowered, the pressure liquid forcing the piston
*h* downward, and this continues until the valve *m*
has reached its central position.   In addition to this

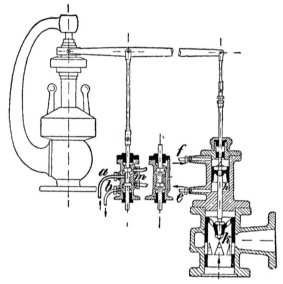

Fig. 78.—Governor of Zölly Turbine.

there is a safety governor, consisting of two weights,
which comes into action when the speed increases
by a certain amount (about 15 per cent) or the main
governor fails, the safety device then displacing a
pin which releases a pawl, and thus closes the steam
valve by means of spring pressure.

The governor operating by throttling the steam is
able to maintain the speed nearly constant, even
under considerable fluctuations of load.

A very similar governor is fitted to the Sulzer turbines. In this case a series of special nozzles, receiving steam from the over-load valve, is provided for dealing with occasional overloads.

(d) Rateau also uses a centrifugal governor—frequently a spring governor—operated through worm gearing from the turbine shaft, a balanced duplex valve remaining at rest during periods of equal output. Furthermore, when the governor is in its extreme positions, a supplementary governing is effected by masking the guide vanes of the turbine in the case of small loads; whilst under higher loads auxiliary steam is admitted to the low-pressure stages of the turbine through a valve on an overload pipe, which of course is uneconomical, whilst under normal output the exhaust steam alone is used for the lower stages.

The governor (Fig. 79) actuates a piston valve $k$ which regulates the admission to the turbine, after the steam has passed through a double-seated valve.

An auxiliary governor, mounted on the shaft of the main governor, releases the double-seated valve through a train of levers, whereupon the valve is closed by a spring (or by hand) as soon as the speed is increased to any considerable degree above the normal.

Rateau turbines are frequently equipped with a compensating device, which returns the governor to its normal position the instant the speed becomes constant, so that, with sufficient throttling, the turbine is restored to its normal velocity.

(e) The Curtis turbine is governed by shutting off separate nozzles in the first-pressure stage by means of valves operated by a centrifugal governor at the upper end of the vertical shaft, and not by throttling.

For this purpose several such expansion valves

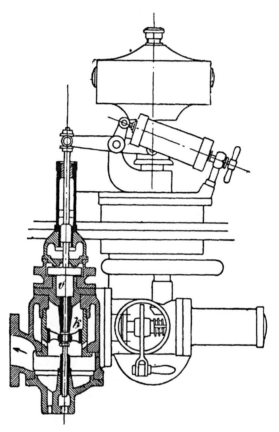

FIG. 79 —Governor of Rateau Turbine.

are set in motion by special pistons mounted on the same ¦ rod as the valves. The upper sides of the pistons are subjected to the steam pressure existing in front of the nozzles, whilst the lower faces of the pistons are under constant pressure from the steam pipe.

If the speed of the turbine increases, the governor causes the piston to close the valve corresponding to its special group of nozzles. In addition, the throttle valve in the live steam pipe can be closed by an auxiliary governor. In earlier patterns the pistons were operated electro-magnetically, but mechanical means are now employed as being more reliable.

To prevent racing, the newer turbines are fitted with an incompletely balanced ring, which is held concentric with the shaft by spiral springs. When the permitted limit of speed is exceeded, the centrifugal force of the unbalanced ring forces it into an eccentric position and causes it to strike against a lever which releases the shut-off valve.

(*f*) The governing mechanism of the Hamilton-Holzwarth turbine is diagrammatically illustrated in Fig. 80. The governor is mounted on an extension of the turbine shaft, and operates the valve by means of a separate device.

The governor weights *mm* are actuated by two discs *ss* on the main shaft, and have, in flying outward, to overcome the force of the springs *ff*. The governing device, the sleeve of which is marked *h*, is operated by the weights *m*, and actuates the spindle *p*, which in turn operates the throttle valve. The displacement of the sleeve of the governor displaces the friction wheel *r* by means of the lever *k* on the shaft *o*, but the said friction wheel cannot turn unless the disc *m* on the solid shaft *q* is displaced to a certain degree by the lever *i* and pressed against the wheel *r*. The friction wheel *r* is turned by the coupling *l* of the hollow shaft *t*, which in turn is driven from the turbine shaft by worm gearing. As

9

the shaft $o$ turns, the valve spindle $p$ is rotated by
the pair of bevel wheels $xy$, the movement in one or
the other direction being slower or quicker according

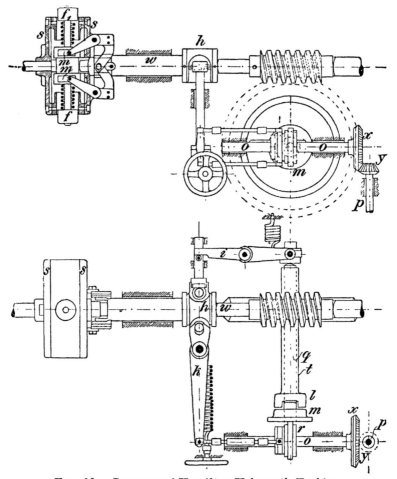

Fig. 80.—Governor of Hamilton-Holzwarth Turbine.

as the wheel $r$ is displaced on the face of the wheel
$m$, this movement of displacement altering both the
speed and direction of the friction wheel. No special
safety governor is needed. Moreover, the speed can

be varied by about 5 per cent while the turbine is running, by means of a double-seated valve between the main shut-off valve and the turbine cover. Steam may be admitted direct to the low-pressure turbine through an overload valve of similar construction, mounted below the governor valve.

(g) The A.E.G. turbines are fitted with a spring governor driven from the main shaft, and acting on an auxiliary piston operated by oil under pressure. The actual governor valve, which takes the form of a throttle valve, is mounted directly behind the shut-off valve. The device is very sensitive, the speed changing by about 2 per cent under a 25 per cent variation in the load, and by not more than 5 per cent when the load is removed entirely.

In smaller machines the governing is effected by a steel band which is rolled up on the inner periphery of a hollow cylinder, and covers or uncovers the nozzles in the cylinder wall. The spring governor on the end of the turbine shaft is similar to that of De Laval.

In the event of the speed increasing beyond about 15 per cent, a rapid closing valve operated by a safety governor comes into action.

(h) In the Parsons turbine gust governing is employed, and (Fig. 81) the steam, after passing through the main stop valve v, enters a balanced governing valve w, suspended on the rod of a piston k in the cylinder e, and pressed toward its seat by the action of the spring f and the piston k. The steam pressure acting underneath the piston k overcomes the force of the spring, so that steam is admitted to the turbine. The space under the piston is in communi-

cation with a valve-gear cylinder, in which a piston
*s* is moved up and down from the shaft of a rotary
oil-pump or from the governor spindle by eccentrics.
During the upstroke the exhaust chamber is placed
in communication with the chamber below the piston,
and the governing valve closes; whilst during the

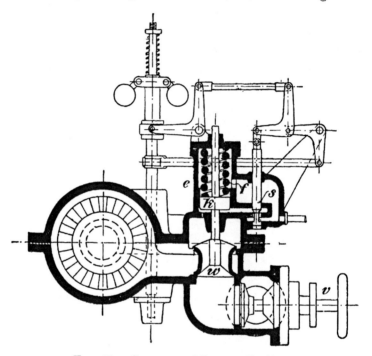

Fig. 81.—Governor of Parsons Turbine.

downstroke the escape of the steam is checked, the
piston *k* being thereby raised and the governing
valve opened.

The governor adjusts the middle position of the
piston valve *s*, thus varying the time at which the
exhaust valve is closed, or varying the aperture of
said valve. If the piston valve makes only a short

stroke, then the valve $w$ is moved farther from its
seat, since the middle position of the piston is lower,
but the valve remains open longer and therefore more
steam is admitted. With a longer piston stroke, and
consequently higher middle position of the piston, the
steam escapes quickly from underneath, so that the
valve opens only to a slight extent and at longer in-

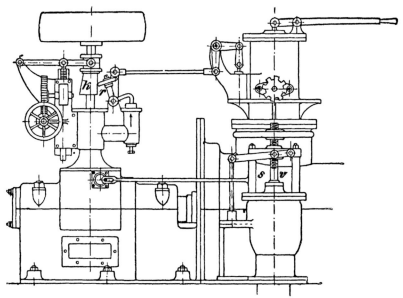

Fig. 82.—Governor of Parsons Turbine, with Safety Valve.

tervals. As the steam only enters the turbine intermittently, about 200 to 500 times a minute, a more or
less extensive throttling takes place.

A somewhat modified arrangement is shown in
Fig. 82, where the governor effects the adjustment
and rocking motion of the valve piston. The piston
has a variable stroke and returns to the closing
position every time. For this purpose the roller $r$
is pressed by a spring against the one-sided cone $k$,

the governing for various speeds of the turbine being adjusted automatically by means of an electrical device. The governor is mounted either on one side of the turbine, between the main-shaft bearing and the collar-thrust bearing, or in the same position in the central plane of the turbine. The governing valve is adapted to be opened by hand in starting the engine. A continuously rotating auxiliary governor in the gearcase of the main governor shuts the steam off in the event of a sudden removal of the load, a double lever causing the valve spindle $v$ to descend, and thus shut off the steam, a pillar $s$ on which the lever is supported by means of cams then being turned.

An arrangement similar to that of Parsons is employed in the Westinghouse turbines (Fig. 83), and needs no further explanation.

Owing to their small and invariable resistance to movement, these governors are able to follow any change in the velocity of the turbine with ease. As a rule the turbine regains its normal speed in three to four seconds after the alteration in the load.

($i$) **Melms & Pfenniger governor.**—This is a shaft governor, the weights of which control an eccentric on a shaft at right angles to the main shaft, the rod of this eccentric operating a lever, and through this a small rotary valve. Above the admission valve is a small governing piston under which the steam is admitted ; and the requisite aperture of the admission valve is secured in accordance with the extent to which the rotary valve is opened.

($k$) In place of the piston valve (Figs. 81, 83), some turbines are fitted with a Rider valve, which is rotated by the governor sleeve, and is displaced along the spindle of the governor by an eccentric. When the speed increases the Rider valve lets the steam out into the exhaust.

(*l*) **Governors acting upon nozzles.**—In the nozzles of each pressure stage are tightly closed by flexible steel bands, and can be opened to a smaller or larger extent by winding or unwinding the toothed bobbins on which the bands are wound. This opera-

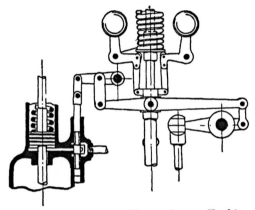

Fɪɢ. 83.—Governor of Westinghouse Turbine.

tion is effected by means of a shaft which passes right through the casing of the turbine, is connected with the governor, and is provided with operating gear for each rotor.

### Thrust-Absorbing Devices.

To take up the axial thrust in reaction turbines *dummy pistons* have to be provided on the main shaft, the total area of which pistons must be equal to the surfaces exposed to steam pressure in the turbine. As a rule the Parsons and Westinghouse turbines are fitted at the high-pressure end with three such pistons, corresponding to the different diameters of the casing (Figs. 38, 39), and the faces of these pistons are placed in communication with the corresponding pressure stages by means of pipes passing through stuffing boxes, or by means of passages made in casting the casing. The pressure between the first stage and the corresponding reliev- ing piston is compensated in the admission chamber, whilst the largest piston corresponding to the final pressure stage is subjected, on the side facing the interior of the turbine, to the steam pressure of the final stage, and on its opposite side to the pressure of the condenser.

The dummy pistons are ground into the casing so as to fit closely but without friction, and they are made steam-tight by labyrinth packing.

In the Melms & Pfenniger turbine, which is a combination of an impulse and reaction turbine, the dummy pistons are dispensed with, the axial pressure from the reaction turbine being taken up by an annular surface arranged between the impulse and the reaction turbines.

### Overload Valves.

In all systems of multiple-stage turbine a device is provided for enabling live steam to be introduced through an overload or by-pass valve (operated by hand or by the governor) direct into one or other of the later stages, when the engine has to deal with a higher than the normal load, so that the working pressure and total output at this stage may be in- creased.

It is true that the opening of this valve impairs

the utilization of the energy of steam and lowers the working efficiency; but these considerations are unimportant in comparison with the resulting advantages, namely, that the turbine can be run, under normal load, at nearly the full boiler pressure in front of the first stage, whereas, otherwise, a considerable degree of throttling would be unavoidable. Moreover, the sphere of utility of the engine is largely increased—frequently more than doubled —though more steam is consumed.

In order to prevent any back pressure on the steam from the high-pressure turbine on opening the by-pass valve, it is advisable that this valve should act like an injector, so that the low-pressure steam is drawn in through the guide vanes.

Like the main valves for ordinary working the auxiliary valves are mostly of the balanced, double-seated type.

The main valve is in operation under all loads and adapts itself to their dimensions, whilst the auxiliary valve comes into action at abnormal speeds only.

### REVERSING.

In the case of marine engines the possibility of reversing the turbine is an important matter. Mounting the vanes of the rotors in a special manner for both directions of rotation has not proved successful in practice, because it is impossible except at the expense of efficiency in steaming ahead—which is the chief consideration in most cases.

Parsons has therefore built separate turbines for running astern, these being mounted in the rear of and on the same shaft as the low-pressure turbines, or in the same casing with the latter (Fig. 84). In running astern the steam then enters the low-pressure turbine, whereas, in steaming ahead, this latter runs empty and *in vacuo*.

After closing the admission to the ahead turbine, the steam can be admitted direct into the astern turbine, which thus acts as a brake in the first instance ; and the more powerful this action, the sooner will the change in the direction of movement of the vessel be accomplished.

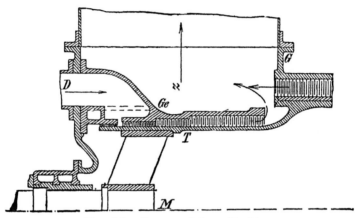

Fɪɢ. 84.—Parsons Astern Turbine.

Z = To condenser.  D = Steam for reversing.  G = Casing of main turbine.  Ge = Casing of astern turbine.  T = Drum of astern turbine.  M = Shaft centres.

Fig. 85 gives a diagrammatical representation of the method of reversing in the case of a triple screw steamer, with high-pressure turbines in the middle, and the low-pressure and reversing turbines on either side. Here *g* is the main regulating valve, whilst *hh* represent the devices for admitting steam into the pipes *l* or *m* through the valves *k*. The vessel will steam ahead when the valve *g* is opened and the

valve $k$ closed; for manœuvring, the valve $g$ is
closed and the valve $k$ opened; and for running
astern steam is admitted into the pipe $m$ by means
of the devices $h$.    In both the latter cases the high-
pressure turbine is shut off entirely, no steam being
admitted.

To accelerate the operation of reversing, large quantities of steam must be admitted into the reversing

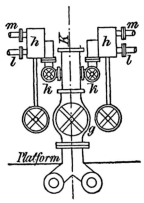

Fig. 85.—Diagram of Reversing Valves.    (K = Live steam.)

turbine; and since the steam must first absorb the
momentum of the rotating parts, the capacity of the
reversing turbine must be at least half as great as that
of the main turbine.    Thus, for instance, in an 8000
H.P. plant it is stated that the main turbine is
brought to a standstill in 69 seconds, and the vessel
is able to run astern at a speed of 13 knots, as
compared with 19 knots when steaming ahead.

Similar reversing turbines have been constructed
by Rateau.

# CHAPTER IV.

## GENERAL ARRANGEMENT OF VARIOUS TURBINES IN PRACTICE.

### 1. The De Laval Steam Turbine (Figs. 86, 87).

THE present type of De Laval turbine is illustrated in Fig. 86.

It is an axial, single-stage, impulse turbine with partial admission, the steam entering at the side and traversing the vanes in a direction parallel to the axis.

The De Laval turbine can be run with steam superheated to the highest practicable temperatures, the consumption of steam and heat decreasing, and the brake horse-power increasing accordingly.

In the drawing, $a$ indicates the turbine shaft, $b$ the turbine disc, $c$ the gear pinion, and $d$, $e$, the bearings.

The gear pinion $c$ drives the larger gear wheels $f$ on either side, the shafts of which are connected with the dynamo shafts by couplings $k$. In addition to the packed bearing $d$, the turbine shaft passes through two bearings in the casing of the gear wheels.

In place of direct coupling, belts or ropes may be used for driving the dynamos, and in such case the driving pulleys are offset on the two parallel shafts.

(140)

Turbines intended to work with and without con-

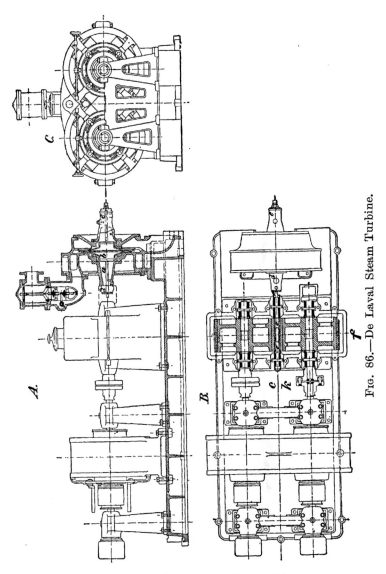

Fig. 86.—De Laval Steam Turbine.

densation alternately are provided with a separate
nozzle ring for each system.

Fig. 87 shows the regulating valve, a balanced duplex valve operated by the governor.

The best theoretical peripheral velocity for the turbine wheel is 950 feet per second with a steam velocity of about 2000, and 2100 feet per second with a steam velocity of 4400 feet per second. The actual peripheral velocity of turbines already

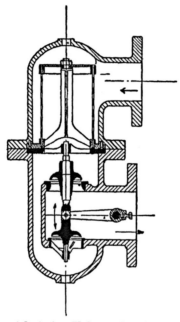

Fɪɢ. 87.—Admission Valve of De Laval Turbine.

built varies between 3690 feet per second in the large sizes and 500 feet per second in the smaller ones. The speeds range between about 10,600 revolutions for the large outputs, about 350 H.P., and up to 30,000 revolutions for the smallest outputs.

The number of steam nozzles is from one to fifteen, according to the size of the turbine. The angle made by the nozzles with the central plane of the wheel should be as small as possible, and the same applies to the vanes on the exhaust side. The nozzle angle is about 18 to 20° for all sizes of turbines. The angle of the vanes is identical on both sides, and is about

Fig. 88.—De Laval Turbine.

36° for large turbines, and 32° for small ones. Even a slight wearing away (about $\frac{1}{16}$ inch) at the edges of the vanes increases the steam consumption by nearly 5 per cent.

With a steam velocity of about 1150 feet per second the radial stresses due to the centrifugal force amount to approximately 22,700 lb. per sq. inch.

When the turbine is running the velocity of the steam varies owing to the fluctuations in steam pressure before reaching the nozzles, the irregular counter-pressure, and according as saturated or superheated steam is used.

The lowest practical admissible steam velocity is 2030 feet, with a steam pressure of 44 lb. per sq. inch, and a counter-pressure of $14\frac{1}{4}$ lb. per sq. inch in the casing. The maximum steam velocity is 4400 feet with a steam pressure of about 200 lb. per sq. inch, a counter-pressure of $\frac{3}{4}$ lb. per sq. inch and 169° F. of superheat.

On account of the high speed the turbines must be geared down 1 : 10 to 1 : 14 between the main shaft and the dynamo.

The speed-reducing gear causes a loss of 1 to 2 per cent in power, is very costly and is also noisy in running.

Helicoid pinions of somewhat mild steel, low in carbon are used ; they are very broad and with small pitch ; their peripheral velocity is about 100 to 165 feet per second and they run in an oil-bath casing. Bronze has been found less satisfactory. All the other running parts are made of nickel steel when superheated steam is used. The shafts are also of (high-carbon) nickel steel, or else of malleable steel.

In the sizes up to about 30 H.P. a single lateral intermediate shaft is used, but the larger sizes are provided with two symmetrical intermediate shafts as shown in Fig. 86.

The loss of energy by the steam in flowing through the vanes is fairly large (up to 30 per cent of the total energy) and the total losses may be estimated at about 40 per cent.

## 2. Rateau Turbine (Fig. 89).

The characteristic features of this turbine are a moderate number of pressure stages separated by packing joints on the shaft, and a moderate number of large rotors with partial admission for the high-pressure stages and full admission for the low-pressure ones.

Since the fall in pressure is fairly small, there is no need for divergent nozzles in the guide wheels, circular orifices being sufficient. Whilst the moderate number of stages results in diminished peripheral velocity, the losses by friction are considerable (10 to 12 per cent), owing to the rotation of the comparatively large rotors in steam of high density.

This turbine shares with all others of the multiple-stage type the advantage of enabling a larger play to be allowed between the guide wheels and rotors. Each of the latter runs in a chamber under constant steam pressure, so that there is no longitudinal thrust, and partial admission is rendered possible.

Fig. 89 illustrates the chief features of the pattern in which a number of rotors are housed in one casing. In this drawing $a$ represents the main steam pipe, $d$ the rotors, $e$ the guide vanes in the fixed partition walls, $g$ the exhaust pipe, $h$ the bearings, and $l$ the overload valve.

A Rateau compound turbine is shown in Fig. 33. This is built in two parts: one for high-pressure

10

steam and the other for low-pressure steam.  The low-
pressure turbine may be fed with steam either direct

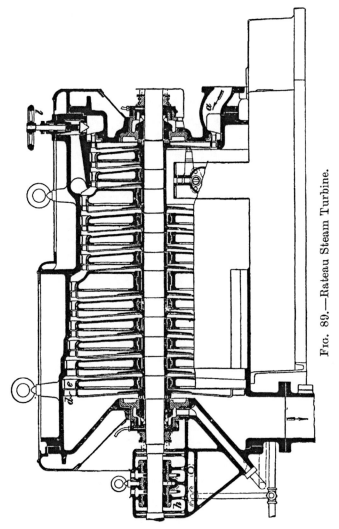

FIG. 89.—Rateau Steam Turbine.

from the high-pressure turbine, or else from a heat accumulator, the working efficiency remaining unaltered.

High-pressure steam is admitted automatically by

the aid of a special governor, as soon as the pressure in the receiver drops below a certain limit.

DIMENSIONS FOR DIFFERENT OUTPUTS.

*Kilowatts.*

| | 100 | 500 | 1000 | 1500 | 2000 | 2500 | 3000 | 4000 | |
|---|---|---|---|---|---|---|---|---|---|
| No. of revs. per min. | 3000 | 3000 | 1500 | 1500 | 1500 | 1000 | 1000 | 1000 | |
| Max. length abt. | 16½ | 21 | 23½ | 27½ | 31 | 33 | 37½ | 44 | ft. |
| ,, width ,, | 3½ | 6 | 7 | 7¼ | 8¼ | 9 | 9¾ | 11½ | ,, |
| ,, height from floor | 3 | 4¾ | 5½ | 6 | 6½ | 7¼ | 7¾ | 9 | ,, |

## 3. The Zölly Turbine (Figs. 90, 91).

This is a multiple-stage impulse turbine, with up to fifteen pressure stages, but without velocity stages. In comparison with the Rateau turbine it has a smaller number of rotors, and therefore fewer pressure stages; and it works with low absolute steam velocities—less than 1300 feet. In other respects the chief differences are of a constructional nature.

It is made with progressive partial admission, and with separate high and low-pressure parts, both, however, being housed in the one casing.

Owing to the small number of pressure stages a comparatively high peripheral velocity is obtainable. Nevertheless the number of stages is large enough to enable divergent nozzles for the several rotors to be dispensed with, simple guides with parallel passages being sufficient.

A characteristic feature is the radial widening of the rotor vanes, this giving a delivery angle smaller than the entering angle. Owing to the outwardly increasing sectional area between the vanes of the rotors, the steam expands, and although there are no expansion nozzles the Zölly turbine must be regarded

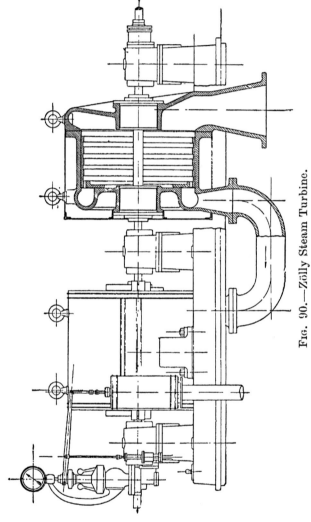

FIG. 90.—Zölly Steam Turbine.

as a multiple-stage impulse turbine without velocity staging.

The governor is operated from the high-pressure

end of the turbine, and tubular coolers for the lubricating oil are provided in the hollow base-plate.

Fig. 91.—Zölly Turbine.

Overloads up to about 20 per cent can be dealt

with when running as a condensing turbine, and about normal output when exhausting at atmospheric pressure can be attained, by admitting steam into the third-pressure stage by means of a by-pass valve.

The steam consumption is approximately :—

| H.-p. | . | . | . | 700 | 1000 | 1800 | 2600 |
|---|---|---|---|---|---|---|---|
| Steam consumption | | | | | | | |
| lb. per h.-p. hour | | | | 12·9 | 13·8 | 12·4 | 11·6 |

The high and low-pressure turbines are mounted in separate casings on a common base-plate, and the bearings are quite independent of the casing.

For low speeds rigid shafts are used, but flexible shafts are preferred for high speeds.

All the parts are illustrated in Figs. 90, 91.

### 4. The Curtis Turbine (Fig. 92).

This vertical turbine is of the impulse type with two to four pressure stages each of which has two to four velocity stages, each rotor accordingly running in a separate chamber. The nozzles in the partition wall of each pressure stage consist of two groups, each forming a separate casting. The admission side of each nozzle is of circular section, that of the delivery side being rectangular. When two pressure stages are employed the vanes of the rotor are bolted laterally on the wheel (Fig. 58), whilst between the vanes of one and the same pressure stage the guide with its guide vanes is bolted on to the casing, these guide vanes reversing the direction of flow for the two velocity stages of the one pressure stage.

The steam is admitted into the first chamber

through two symmetrically mounted annular slits in

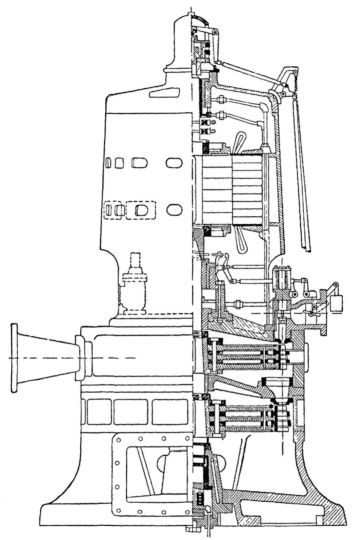

Fig. 92.—Curtis Steam Turbine.

the cover, the valve chests with the admission nozzles being arranged above them.   Governing is effected, not by throttling, but by modifying the volume of steam admitted, the nozzle valves being either fully opened or completely shut.

The degree of admission in the several pressure stages increases with the expansion of the steam.

The amount of play left between the rotor vanes and the guide vanes is as follows :—

| Output. | One. | Two. | Three. | Four or Five Stage. |
|---|---|---|---|---|
| 500   kilowatts | 0·1 | 0·1 | 0·1 | 0·15 inch. |
| 8000      ,, | 0·2 | 0·2 | 0·2 | 0·33   ,, |

The nozzles of the first-pressure stages arranged in the cover can be closed separately by valves, operated independently by special adjustment devices, and to which steam is admitted through spindle valves separately influenced by the governor. In addition, an auxiliary throttling can be effected by means of a safety governor.

The nozzles of the second stage may also, in some patterns, be closed by the aid of slide valves that are adjustable from outside.

Small units, up to 300 kilowatts, are built with horizontal shafts.

The weight of the entire turbine is supported by the bottom footstep bearing, which is lubricated with oil or water under pressure. The oil pressure amounts to as much as 710 lb. per sq. inch, and the oil consumption is about 1 pint per minute.

Some of the newer Curtis turbines run with a peripheral velocity of 450 feet.

## 5. A.E.G. Turbines (Figs. 93, 94).

Recognizing that the employment of velocity stages affords certain special advantages over pressure staging pure and simple, inasmuch as it protects the

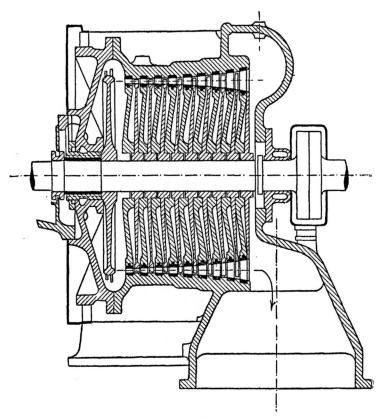

Fɪɢ. 93.—A.E.G. Steam Turbine.

casing from direct contact with steam at high temperature and pressure, whilst purely pressure stages are needed in the low-pressure end for large outputs, the A.E.G. makes a number of different patterns of turbine.

In the smallest models one pressure stage is provided with three velocity stages, the rotor runs freely

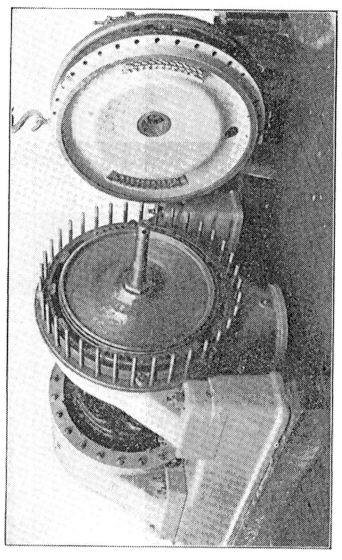

FIG. 94.—A.E.G. Turbine.

and the admission nozzles are arranged in the detachable cover, the diameter of the vane rim being

large enough to utilize the kinetic energy completely, by a sufficiently high peripheral velocity.

In a condensing turbine running at about 3000 revolutions per minute, two pressure stages, each with two velocity stages, are used. The rotors are situated in two chambers, separated by partition walls, and the end of the shaft is carried by a bearing cast on to the casing cover, this bearing also taking the weight of the governor, which is mounted on the end of the shaft.

In plants with a poor vacuum, the low-pressure wheel has only one pressure stage.

To enable a considerable fall in pressure to be utilized at once in the first stage, another pattern is provided with a far larger high-pressure wheel, with two velocity stages, in contrast to the low-pressure wheels in the same casing, these latter wheels being separated from each other by partition walls and thus acting as separate impulse turbines, their guide vane rims serving to generate velocity from the fall in pressure still available.

For outputs up to 3000 kilowatts for instance, one high-pressure rotor, with two velocity stages, and nine low-pressure rotors are used.

## 6. The Hamilton-Holzwarth Turbine (Fig. 36).

This is a purely multiple-stage impulse turbine with full admission to the rotors in which the increase in velocity and the expansion of the steam take place. Each pair of fixed guide wheels forms a chamber housing one rotor.

For outputs about 750 kilowatts the high and low-pressure turbines are housed in separate casings, and in order to keep down the number of rotors the number of stages in the low-pressure side is increased, so that the velocity of the steam increases accordingly. The governor is mounted direct on the shaft. To enable overloads to be dealt with, live steam can be introduced into the low-pressure turbine by means of an auxiliary nozzle which acts like an injector.

The various parts have been described in earlier chapters. The several portions of the shaft are connected together by flexible couplings.

### 7. Elektra Radial Impulse Turbine (Fig. 95).

This turbine is built by the Gesellschaft für Elektrische Industrie. In this instance the high peripheral velocities of the De Laval turbine are obviated by means of repeated admission to the rotor (see p. 26), so that with a maximum velocity of about 270 feet the stresses on the material are only moderate. The turbine is governed by adjusting the nozzle area by means of flexible tongues controlled by a shaft governor, or else by acting on the throttle valve.

In smaller sizes the turbine is built with one rotor, one pressure stage, and four velocity stages, whilst in larger sizes there are two pressure stages, each with three velocity stages, by means of which arrangement a steep fall in pressure with moderate peripheral velocity can be dealt with.

The rotor is provided on its end surface with steel vanes, and its hub makes a tight joint against the casing $g$ by means of labyrinth packing. The casing is surrounded by an annular channel $k$, into which the steam enters through a pipe $a$, and passes thence into the rotor by way of the nozzles $dd$. After tra-

versing several vanes the steam is returned to the
rotor by the passage $u_1$, flows inwardly through the

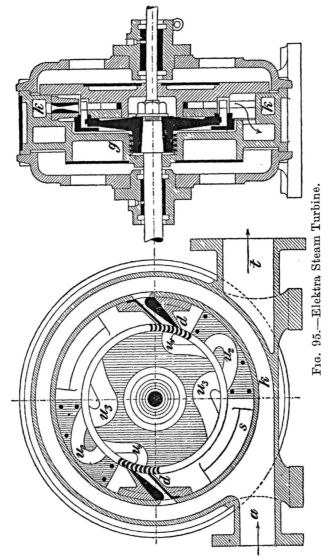

FIG. 95.—Elektra Steam Turbine.

outer passage $u_2$, traverses the turbine vanes again
and is admitted a third time through the passage $u_3$.

The sectional areas of the passages are of such dimensions that the steam traversing the vanes has lost all its velocity by the time it issues from $u_3$, and escapes through the exhaust $s$, and pipe $t$. Hence there is a fourfold admission in the direction of rotation of the wheel.

The rotor is slipped on to a conical part of the shaft, and is held in position by a nut.

For an output of about 50 H.P. the rotor has a diameter of about 20 inches and a speed of 3000 revolutions. The number of vanes in this case is 400. The steam consumption about 16 lb. per B.H.P. hour for an output of 200 H.P. and about 20 lb. for 56 H.P. The turbine is built both for direct coupling with a dynamo and for belt driving.

Another radial admission turbine is the Backström-Smith engine, a multiple-stage impulse turbine, with one velocity stage in each pressure stage, and arranged for partial admission.

## 8. Parsons Turbine (Figs. 96, 97).

The characteristic feature of the Parsons turbine is moderate diameter coupled with relatively great length, and a large number of pressure stages for slight falls in pressure. In the case of very large outputs the length can be diminished by increasing the diameter of the drum and reducing the number of the stages. When the drum diameter is very large, the high and low-pressure members are housed in one cylinder.

The fact that the drum is generally made in three steps is solely on account of constructional considerations. The velocity of the steam can be modified by

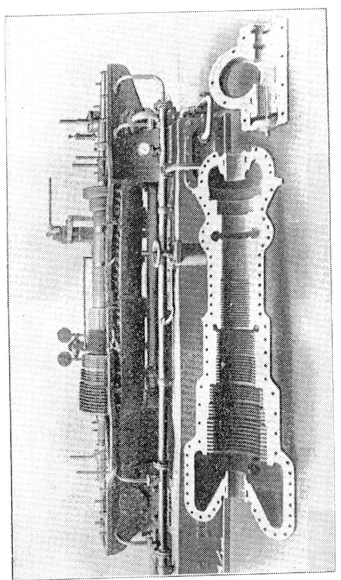

FIG. 96.—Parsons Turbine with Cover removed.

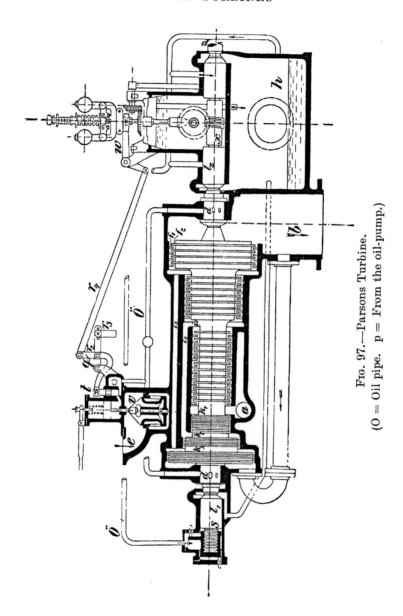

Fig. 97.—Parsons Turbine.

(O = Oil pipe.  p = From the oil-pump.)

altering the diameter of the drum, so that practically suitable lengths of vane can be obtained at both the high and low-pressure ends.

The advantage of the large number of stages (120 to 160) is that the steam is enabled to give out the whole of its energy as completely as possible, even where the conditions of velocity are unfavourable. Since a fall in pressure takes place both in the guide vanes and rotor vanes, the velocity of the steam increases with relation to that of the vanes. The amount of play between the rotor and the casing in the high-pressure side is $\frac{1}{32}$ to $\frac{1}{25}$ of an inch, and on the low-pressure side $\frac{1}{8}$ to $\frac{1}{5}$ of an inch. There is no need for any small axial play between the guide vanes and those of the rotor, and the radial play should be as small as possible, in order to prevent any leakage of steam between one stage and another.

The greatest loss on this score occurs at the high-pressure end, the radial areas allowing such loss at the low-pressure end being small in relation to the volume of steam. The losses are minimized by the fact that the velocity of the steam that escapes without doing work is lowered by the frictional resistance against the casing, by the compression to which the steam is subjected by centrifugal force, and by the continuous change of direction in the guide vanes.

For capacities up to 1000 H.P. three drums are employed, the stages are usually eight in number, and each stage comprises five to ten rotors. In larger sizes the number of vanes and also of turbine rims increases. The total vane area in a 1000 H.P.

11

turbine is about 270 sq. feet, and the diameter of the drum ranges from 15 inches to 3 feet.

The Parsons turbine is not suitable for small outputs, owing to its very complicated build and the consequent excessive clearance losses.

The number of revolutions per minute in different sizes is given below :—

|  | 100 | 1000 | 1500 | 4000 | 5000 | 10,000 H.P. |
|---|---|---|---|---|---|---|
| $n =$ | 3500-5000 | 1500-2500 | 1000-1200 | 1000-1200 | 1000-1200 | 750-1000 |

With still larger sizes the number of revolutions decreases considerably, being only 160 per minute in the 68,000 H.P. turbines of the Cunard liners (two high and two low-pressure turbines, each of 17,000 H.P.).

The peripheral velocity is 195 to 330 feet.

Fig. 97 shows the general arrangement of the Parsons turbine, whilst Fig. 104 represents the marine pattern, in which the dummy pistons are dispensed with. In these Figs. $e$ indicates the admission, $b$ the exhaust, $f_1$ the guide vanes on the casing, $f_2$ the guide vanes on the three-step drum, $k1$, $k2$, $k3$ the dummy pistons for the axial thrust, $i_1$, $i_2$, $i_3$ the communicating passages between the steps of the drum and the dummy pistons, $s$ the collar-thrust bearing for adjusting the guide vanes in relation to the rotor vanes, $d$ stuffing boxes, $l_1$, $l_2$, bearings, $x$ the worm driving the governor shaft and the oil-pump in the oil reservoir, $v$ admission valve, $t$ piston valve operated from the governor shaft by the lever $q$, and the rods $r_2$, $r_3$, $r_4$, $w$.

The several parts have already been described in earlier chapters.

Of very similar build to the Parsons turbine is the Allis-Chalmers turbine.

### 9. Westinghouse Turbine.

The Westinghouse Co. which formerly built chiefly Parsons turbines (Fig. 98), now manufactures a pattern which is a combination of impulse and reaction turbine.   In the first stage is an impulse wheel with double velocity staging and an admission consisting of three nozzles.   Thence the steam passes to the second stage, composed of a number of Parsons rotors.   After traversing this stage the current of steam divides, the one portion taking its course through several Parsons turbines, whilst the other is conducted to the opposite end of the turbine, where it also acts on a number of Parsons wheels. In this way a compensation of pressure is obtained without the necessity for employing long dummy pistons.

### 10. Melms & Pfenniger Turbine.

This also is a combination of impulse and reaction turbines, and both the high and low-pressure vanes are secured on drums.   Since the high-pressure side works with partial admission, the drums are of very large diameter, whereas those on the low-pressure side are smaller, and the end ring surface at the point of transition between them can be utilized as a balance surface, thus reducing the total length of the drum (see Fig. 42).

### 11. Sulzer Turbine.

The new Sulzer turbine (Fig. 40) is similar to the foregoing, an impulse turbine with velocity staging serving as high-pressure stage, and when the steam has so far expanded that its volume is sufficient for

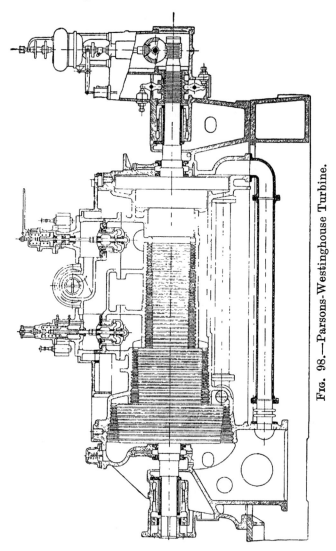

FIG. 98.—Parsons-Westinghouse Turbine.

full admission to a rim of suitable diameter, it traverses the vanes of a reaction turbine, these vanes being secured on stepped drums.   The conical nozzles are of milled steel.

## 12. Kerr Turbine.

In this type rotors with Pelton vanes are used, the nozzles forming a pressure stage.   When several pressure stages are employed, a corresponding number of rotors revolve in separate compartments of the casing, each of which is equipped with nozzles.

# CHAPTER V.

## 1. Degree of Vacuum and Arrangement of Condenser.

SINCE the expansion of the steam is carried much farther in turbines than it is in reciprocating engines (140 to 160 times, as compared with 14 times), every reduction that can be made in the back pressure from the condenser is accompanied by an increase in efficiency. In all turbine systems the vacuum produced in the condenser by the vacuum pump is independent of the operations within the turbine itself, and is controlled entirely by the temperature of the cooling water and size of the vacuum pump. Moreover, every reduction of pressure in the condenser of the steam turbine means a far larger amount of saved energy than in the case of the reciprocating engine.

Nevertheless, even the best of condensers cannot develop its maximum efficiency unless the steam is able to expand to condenser pressure before issuing from the last rotor.

In all systems in which the rotors revolve in very dense media, for example in multiple-stage impulse turbines, it may frequently happen that a more perfect vacuum in the condenser becomes necessary.

Such a vacuum is not required in the case of axial

(166)

turbines working with full admission, because the steam is enclosed in a ring, and the entire periphery shares in the production of the rotary movement, so that almost the same effect can be obtained with a less effective vacuum. A reduction of 1 per cent in the efficiency of the vacuum results in an increase of 2 per cent in the steam consumption. In the case of single-stage impulse turbines the efficiency is considerably impaired by a poor vacuum.

Saturated steam and a moderate vacuum can also be utilized in the Parsons turbine; but if the steam becomes wet the velocity declines, and only regains its normal value when the steam has become drier. For steam with 2 per cent of moisture the efficiency diminishes by about 4 per cent.

Non-condensing turbines require 50 to 60 per cent more steam than condensing turbines.

In order to obtain the condensed steam free from oil it is necessary to avoid the use of central condensing plants which receive greasy steam from other engines.

By increasing the vacuum the actual saving effected is three-fourths of the theoretical economy. The power consumed in working the condenser is about $4\frac{1}{2}$ to 2 per cent of the total output, according to the size of the turbine.

The types of condenser used are the same as for reciprocating engines.

With surface condensers about forty to forty-three volumes of cooling water are required, and the condensed water, which is free from oil and has a temperature of about 104° F., can be used as boiler feed. As a rule the steam and water pass through the condenser

in opposite directions, the condensed water being col-
lected in a chamber underneath, whence it is dis-
charged by an air pump.   In all cases it is necessary
that the cooling surfaces should act energetically, and
high-speed air pumps must be used if the size of the
condenser is to be kept down within reasonable limits.
In addition to the usual wet air pump, Parsons uses
for augmenting the vacuum a dry pump and a
special, small auxiliary condenser H (Fig. 99), into

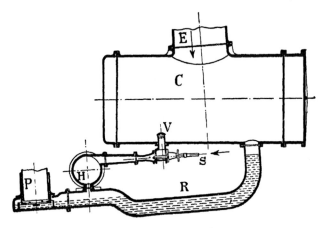

Fig. 99.—Parsons Condenser.

which air and steam are drawn from the main condenser C by means of a steam-operated ejector S.   V is
a non-return valve, P the pump, and R a water seal.
Electromotors are preferably employed for driving the
condenser pumps.

In view of the requisite rapid transference of air
and water, centrifugal pumps offer certain advantages, and arrangements of this kind have been sug-
gested by Stumpf.

Rateau provides ejector condensers, and space is
saved and a good vacuum obtained in the Curtis tur-

bine by the provision of an ejector condenser in the base-plate of the turbine.

In this type of condenser the water drawn in by the vacuum is given the requisite velocity by means of a pump, so that the condensed steam is moved and brought into contact with the external air. The velocity generated in the centrifugal pump is transformed into pressure in the guide apparatus of the pump, and reconverted into velocity in a nozzle.

The Leblanc condenser is worked without a plunger pump, the requisite velocity being imparted to the water by a partial-admission centrifugal pump without guides or casing. The pump and nozzle are in one. In jet-mixing condensers the cooling water is sprayed into a condenser to which the air-injection pumps are attached.

Jet condensers are more suitable for small installations.

### 2. Utilizing Exhaust Steam for Working Turbines—Heat Accumulators.

The exhaust steam from reciprocating engines and turbines—especially when superheated steam is used —still contains large quantities of heat which escape without being utilized. On this account various attempts have been made to give the condensers a form enabling them to serve as evaporators for liquids of very low boiling-point, in order that the resulting vapours may be utilized further.

Thus, for instance, the construction of the De Laval turbine has been modified in such a manner that it is able to utilize the exhaust steam from winding engines, steam hammers, etc., about 65 per cent of

the heat of the exhaust steam being recovered by
this means in the form of work.   It is true that these
turbine plants with low speed are larger and heavier
than high-pressure turbines ; but nevertheless the
results obtained are superior to those from condensation by about 30 per cent.

Any system of turbine may be used, and about
39 to 40 lb. of steam are required to produce 1 kilowatt hour.

Compensating devices (heat accumulators) are
needed to obtain uniform utilization in the case of
rolling-mill engines, winding engines, steam pumps,
presses, shears, hammers, etc., which furnish exhaust
steam intermittently and under a pressure of about $0 \cdot 1$
to $0 \cdot 3$ atmosphere.   The construction of these accumulators, as designed by Rateau, is shown in Fig. 100.
The exhaust steam is fed into the vertical or horizontal
receiver, which is filled with water ; and this steam
is condensed therein during the periods of abundant
supply, re-evaporation taking place during the periods
when the supply of steam declines or ceases.   Correspondingly, the pressure in the receiver increases
while exhaust steam is being admitted, and diminishes
when the low-pressure turbine is consuming steam.

By this means, and with an admission pressure of
$14\frac{1}{4}$ lb. per sq. inch and a back pressure of $1\frac{1}{2}$ lb.,
the low-pressure turbine consumes about 33 lb. of
steam per B.H.P.   In the new form of Rateau receiver (Fig. 100), steam is introduced in the form of
jets into the water vessel $a$, through perforated elliptical tubes $bb$, which are riveted to the bottom of the
vessel, the resulting steam being conveyed direct to
the low-pressure turbine.

The steam consumption of an exhaust-steam tur-

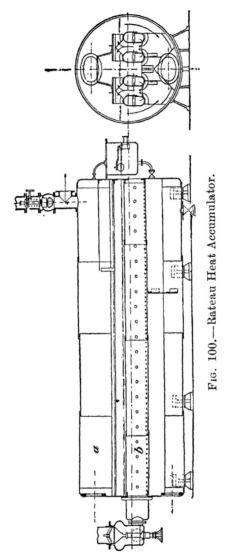

FIG. 100.—Rateau Heat Accumulator.

bine is about double that of a high-pressure turbine of equal dimensions.

So long as the high-pressure engine (for instance,

a winding engine) is running under full load, and gives off more low-pressure steam than the low-pressure turbine can consume, the major portion of the excess condenses in the receiver, which consequently absorbs the heat of same.   During the stoppages of the winding engine, or when the exhaust steam from that engine is insufficient for the needs of the reaction turbine, steam is formed in the receiver.

During prolonged stoppages the turbine can be supplied direct with live steam, this being effected automatically when the pressure in the heat accumulator sinks below a certain limit.   With receivers of sufficient size the temperature of same may be kept constant within a range of $5\frac{1}{2}°$ to 119° F.

In this way, when the principal engine consumes 2000 lb. of steam per hour, about 660 H.P. can be recovered, and about 2200 H.P. from 7000 lb. of exhaust steam.

# CHAPTER VI.

## PRACTICAL APPLICATION OF STEAM TURBINES.

### *Dynamos.*

THE development of the steam turbine has greatly influenced the construction of the dynamo; whilst, conversely, the increase in the speed of the dynamo gave an impetus to the elaboration of a prime mover capable of satisfying that requirement. The fact that the weight of the generator can be reduced when the speed is increased, and heavy flywheels can be dispensed with, pointed to the desirability of increasing the dynamo speed to a limit approximating to the lowest economical speed of the steam turbine; and the growing importance of alternating-current generators greatly favoured the introduction of steam turbines. With the customary periodicity of about fifty per second, the maximum speed of two-pole machines is about 3000 revolutions, and for four-pole dynamos 1500 revolutions. This latter speed may be regarded as the highest limit for machines with an output of more than 1000 kilowatts, in view of the increasing difficulty of balancing the mass of the revolving parts, and the high rubbing speed in the bearings of the dynamo.

Dynamos of the two-phase or poly-phase type are best suited to the peculiarities of the steam turbine, since in this case the difficulties attending the use of commutators at high speeds are eliminated; and besides, dynamos with stationary armature windings and rotary field magnets are the most convenient to build.

Hence, direct-current dynamos are less suitable than alternating-current generators for steam turbine drive.

Moreover, since the speed of the De Laval and Parsons turbines varies by barely 2 per cent between full load and running light, it is easy to couple alternating dynamos driven by reciprocating engines in parallel with those driven by turbines.

Owing to the uniform torque of the turbine, the periodical pulsations of coupled motors—a frequent source of trouble in the co-action of alternating machines—are obviated.

Low-speed dynamos coupled in parallel with turbo-dynamos are far more evenly loaded, the more sensitive governing of the latter taking up the fluctuations of the line load, so that they act as equalizers.

In many instances De Laval turbo-generators have been mounted on the boilers of locomotives—from which they receive steam direct—for the purpose of train lighting.

Attempts have also been made to use turbines for driving motor-cars and torpedoes.

### High-lift Centrifugal Pumps.

Existing systems of turbine have also proved well adapted for driving centrifugal pumps, especially the De Laval and Rateau turbines.

To attain their maximum efficiency for given

FIG. 101.—Parsons Turbo-dynamo.

volumes of water, centrifugal pumps require to be driven at definite velocities. Small De Laval

Fig. 102.—Parsons Turbo-dynamo.

turbines, with a single intermediate shaft, are sufficient to deliver against a head of 16 to 160 feet, and larger sizes, with two shafts, for heads of 40 to 330 feet. For still greater heads the pump may be coupled direct with the shaft of the turbine, the speed then attaining 10,000 to 30,000 revolutions per minute. Nevertheless, in such cases, the vane wheel of the pump is of very small dimensions and will no longer draw, so that it has to be fed by a pump running at a far lower speed, this second pump drawing up the water and delivering it to the high-speed pump. Pumps of this kind are built with a single wheel for delivering against heads up to 980 feet; and when smaller quantities of water have to be raised, an efficiency of 64 per cent is still attained with a head of 1000 feet.

In any case much higher lifts can be obtained by using steam turbines than with any other type of pump, since the pressure set up by the vane wheel of a centrifugal pump increases as the square of the peripheral velocity, and consequently with the velocity generated by the steam turbine.

Thus, for instance, Rateau turbines are employed to drive the pumping plant of a waterworks where nearly 5000 gallons of water are raised per minute against a head of 460 feet, with a steam consumption of 15 lb. per B.H.P. hour, the working pressure being 210 lb. per sq. inch.

### Feed-water Pumps.

Steam turbines also exhibit definite advantages in comparison with ordinary reciprocating methods of driving feed-water pumps, the consumption of steam

12

being only 39 to 66 lb. per brake H.P. against 140 to 220 lb. An essential condition to success, however, is that the units in the groups of boilers should be of approximately equal dimensions.

The feed pumps operated by Rateau turbines are fitted with automatic controlling devices, which are said to act satisfactorily even when the consumption of feed water is reduced to a minimum.

### *Blowers.*

Blowers working at a gauge pressure of over 4 inches can be advantageously driven by De Laval turbines, for instance, their high velocity being highly suitable for turbine work; and the same may also be said about Rateau and Parsons turbines. Thus a 10-inch blower driven by the first-named turbine at a speed of 335 revolutions per second and a water-gauge pressure of 224 inches delivered 1 cubic foot of air, 4 H.P. being required for a working pressure of 145 lb. per sq. inch.

As in the case of centrifugal pumps it is possible, by coupling several blowers together, to deliver air at a pressure of 90 lb. per sq. inch with an expenditure of 350 H.P.

By means of a special system of governing, the pressure can be maintained at a constant level whatever the output.

There is a great future for the employment of air compressors in metallurgical works and mines, because the blowing engines for blast furnaces, steel-works plant, etc., can be of much simpler and smaller construction when centrifugal machines are used, particularly when it is desired to utilize exhaust steam.

*Marine Propulsion.*

In spite of their extensive application, the advantageous use of steam-turbines for marine propulsion is still encountered by difficulties. Up to the present there has been no really unimpeachable comparison between the turbine and the highly perfected multiple-expansion engine, in respect of steam consumption. Except for the "Turbinia," which was experimented with in 1897, the first seagoing ship to be equipped with the Parsons turbine—which at present holds the field for marine propulsion—was built in 1901.

It being necessary to run the turbines at very high peripheral velocities, the dimensions of the propeller—both as regards diameter and blade area—have to be curtailed, the diameter being about one half, and the blade area one-fourth the dimensions usual with reciprocating engines.

In particular, the plant must be arranged differently for mercantile vessels and warships, since the latter are only occasionally run at full speed, whereas this is the normal condition for the merchant ship. Hence, with a smaller engine output, it is hardly possible at present to obtain economical steam consumption in the warship. In addition it must be borne in mind that the employment of high-working pressures is less important in the case of steam turbines than in reciprocating engines, whilst on the other hand a high vacuum in the condenser is an essential condition to the proper utilization of the steam. In any case, however, the use of turbines is the easier in proportion as the vessel is larger.

The usual patterns for merchant vessels have mostly three (Fig. 103) or four shafts, each with a propeller. A high-pressure turbine is installed amidships, and a low-pressure turbine with astern turbine on either side, these receiving steam from the high-pressure turbine and discharging it into the condensers.

Since marine engines have to be reversed in order to enable the vessel to move forward and astern, and experiments in the construction of reversible turbines have met with no practical success, Parsons employs special astern turbines, mounted in the casing and on the shaft of the high- or low-pressure turbines. During the forward movement of the vessel the astern turbines run empty, whilst in running astern the high and low-pressure turbines behave in a similar manner (Fig. 104).

In order to obtain economy of steam consumption in warships, it is usual to employ separate (cruising) turbines for the low speeds constituting the normal; but this entails increased engine space and weight.

With modern reciprocating engines the steam consumption at low speeds is barely half that of the steam turbine.

For a warship with four shafts, of which each of the two central shafts carries a low-pressure main turbine $N_1$ and $N_2$, and a high or low-pressure cruising turbine $Mh$ or $Mn$, whilst each of the port and starboard shafts carries a high-pressure main turbine $H_1$ or $H_2$ and an astern turbine $R_1$ and $R_2$, the steam takes the following course ($\rightarrow$) for forward running.

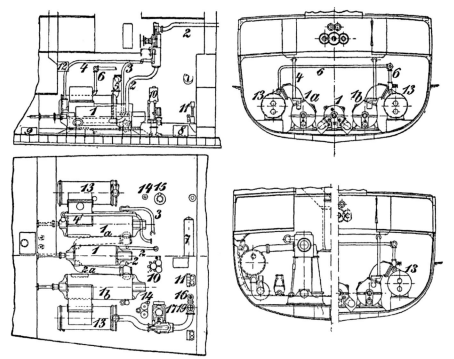

FIG. 103.—Arrangement of Parsons Turbines on a Ship.

| | |
|---|---|
| 1a. High-pressure turbine. | 7. Feed water heater. |
| 1. Intermediate turbine. | 9. Oil tank. |
| 1b. Low-pressure turbine. | 10. Air pumps. |
| 2. Main steam-pipe. | 11. Ballast pumps. |
| 2a. Pipe to low-pressure turbines. | 12. Reserve oil tank. |
| | 13. Condenser. |
| 3. Steam-pipe for forward manœuvring. | 14. Air pumps. |
| | 15. Air pumps. |
| 4. Steam-pipe for manœuvring astern. | 16. Emergency injector. |
| | 17. Centrifugal pumps. |
| 5. Main exhaust. | 19. Cooling water intake. |
| 6. Auxiliary exhaust. | |

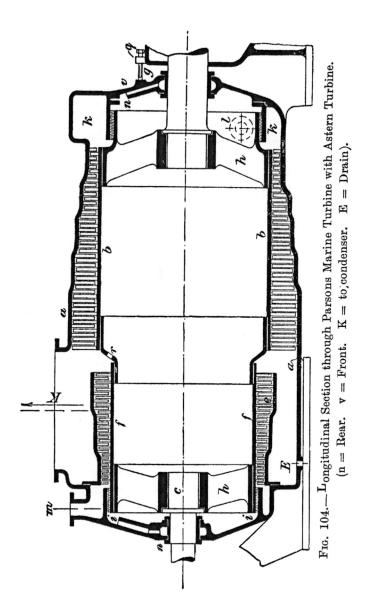

Fig. 104.—Longitudinal Section through Parsons Marine Turbine with Astern Turbine.

(n = Rear.  v = Front.  K = to condenser.  E = Drain).

Heavy output :—

$$\rightarrow H_1 \rightarrow N_1 \rightarrow \text{condenser.}$$
$$\rightarrow H_2 \rightarrow N_2 \rightarrow \text{condenser.}$$

Medium output :—

$$Mn \rightarrow \Big\langle \begin{array}{l} H_1 \rightarrow N_1 \rightarrow \text{condenser.} \\ H_2 \rightarrow N_2 \rightarrow \text{condenser.} \end{array}$$

Minimum output :—

$$\rightarrow Mh \rightarrow Mn \Big\langle \begin{array}{l} H_1 \rightarrow N_1 \rightarrow \text{condenser.} \\ H_2 \rightarrow N_2 \rightarrow \text{condenser.} \end{array}$$

For large outputs the main shut-down valve can be throttled, whilst for small outputs $H_1$ and $H_2$ usually do little or no work.

An essential advantage of the turbine in comparison with the reciprocating engine is the abolition of the troublesome vibration of the vessel; the low height, which is particularly valuable in the case of warships, since it enables the engines to be mounted below the armoured deck; and the non-necessity for heavy foundations. In general, the amount of floor space is about the same as for reciprocating engines; but the heavy thrust blocks required in the latter case to take up the thrust of the propellers are dispensed with, the axial thrust set up in the Parsons turbine by the reaction effect in the direction of the flow of steam being almost sufficient to absorb the propeller thrust, thus rendering special dummy pistons also unnecessary.

Nevertheless, it may be confidently asserted already that the steam consumption at full speed does not work out in favour of the turbine engine. Whether the steam turbine is economical in the case of vessels (cargo boats) which run at only a low speed, is a point that needs further investigation.

Fig. 104 gives a diagrammatical representation of the general arrangement of a Parsons low-pressure turbine and astern turbine for marine work.

In this drawing—

$a$ = the divided casing.

$b$ = the rotary drum.

$c$ = drum spindle.

$e$ = casing of astern turbine.

$f$ = drum of same.

$g$ = stuffing boxes.

$h$ = cast-steel wheel.

$i$ = labyrinth packing.

$k$ = steam intake.

$l$ = steam for forward manœuvring.

$m$ = steam for manœuvring astern.

$n$ = steam for the stuffing boxes.

$q$ = adjusting bolts.

$r$ = passage to condenser for waste steam.

### Floor Space.

Whilst a double expansion engine of 1000 H.P. with a dynamo coupled to the fly-wheel shaft requires a floor space of about 100 sq. inches per H.P., only 25 sq. inches per H.P. are needed for a Parsons turbine; whilst for an output of 4000 H.P. the corresponding figures are 250 sq. inches and 75 sq. inches. The same floor space that would be required for the erection of three vertical reciprocating engines with an aggregate of 900 H.P. will accommodate five Parson turbines, each of 600 H.P. A 150 H.P. horizontal double expansion engine needs a floor space of about 1½ sq. feet per H.P., a Parsons turbine of equal capacity about 3½ sq. inches, and for an output of 7500 H.P., without a dynamo, the space required is about 12 sq. inches. In point of relative weight, too, the steam turbine has the advantage over stationary reciprocating engines, a Parsons turbine weighing 56 to 33 lb. per H.P., according to the size, whereas a corresponding reciprocating engine weighs 350 to 200 lb. per H.P.

The floor space occupied by a De Laval turbine of 3 H.P. is about 1½ sq. inches per H.P. : of 160 H.P. 50 sq. inches per H.P., and for 300 H.P. 55 sq. inches per H.P.

*Relative Advantages of Turbines and Reciprocating Engines.*

Comparing reciprocating engines with steam turbines, it may be said that the former utilize the steam better at high pressures, whilst the turbine has the advantage when low pressures are in question.

For very large powers the steam consumption of the turbine is about equal to that of the modern triple-expansion engine. Otherwise the latter, when condensing, consumes 7 to 8 per cent less steam ; but this advantage is counterbalanced by other valuable properties of the steam turbine, as already pointed out.

[THE END.]

# INDEX.